INDIGENOUS VOICES IN DIGITAL SPACES

Indigenous Voices
in Digital Spaces

CINDY TEKOBBE

UTAH STATE UNIVERSITY PRESS
Logan

© 2024 by University Press of Colorado

Published by Utah State University Press
An imprint of University Press of Colorado
1580 North Logan Street, Suite 660
PMB 39883
Denver, Colorado 80203-1942

All rights reserved
Printed in the United States of America

 The University Press of Colorado is a proud member of Association of University Presses.

The University Press of Colorado is a cooperative publishing enterprise supported, in part, by Adams State University, Colorado State University, Fort Lewis College, Metropolitan State University of Denver, University of Alaska Fairbanks, University of Colorado, University of Denver, University of Northern Colorado, University of Wyoming, Utah State University, and Western Colorado University.

∞ This paper meets the requirements of the ANSI/NISO Z39.48-1992 (Permanence of Paper).

ISBN: 978-1-64642-645-4 (hardcover)
ISBN: 978-1-64642-646-1 (paperback)
ISBN: 978-1-64642-647-8 (ebook)
https://doi.org/10.7330/9781646426478

Library of Congress Cataloging-in-Publication Data

Names: Tekobbe, Cindy, author.
Title: Indigenous voices in digital spaces / Cindy Tekobbe.
Description: Logan : Utah State University Press, [2024] | Includes bibliographical references and index.
Identifiers: LCCN 2024002515 (print) | LCCN 2024002516 (ebook) | ISBN 9781646426454 (hardcover) | ISBN 9781646426461 (paperback) | ISBN 9781646426478 (ebook)
Subjects: LCSH: Veregge, Jeffrey—Interviews. | Choctaw Indians—Communication—Case studies. | Choctaw language—Rhetoric—Case studies. | Choctaw Indians—Ethnic identity—Case studies. | Choctaw Indians—Folklore. | Indigenous peoples—Communication—Case studies. | Indigenous peoples—Ethnic identity—Case studies. | Digital media—Social aspects—Case studies. | Social media—Case studies.
Classification: LCC E99.C8 T44 2024 (print) | LCC E99.C8 (ebook) | DDC 302.23/108997387—dc23/eng/20240314
LC record available at https://lccn.loc.gov/2024002515
LC ebook record available at https://lccn.loc.gov/2024002516

Cover art: © Bree Island/Mixed Creatives, ᐊᏂ ᓯᑊᑯᑉ ati-nîkân ~ in the future. Digital Illustration. 2022.

For Chris
In memory of Helen Jane, my mother

Contents

List of Figures ix

Acknowledgments xi

Preface xv

Introduction: Doing Storytelling as Epistemology 3

1. Indigenous Storytelling and Ways of Thinking and Being 32
2. Listen: Survivance and Decolonialism as Method in Thinking about Digital Activism 52
3. Skoden: Indigenous Identity Construction through Facebook Memes 75
4. Jeffrey Veregge: A Story of Relations 106
5. MazaCoin: Decolonizing a Colonial Fantasy 129

Conclusion 151

Notes 169
References 173
Index 181
About the Author 189

Figures

3.1. "When You See the Invisible Hand" meme   85
3.2. "Netflix Adaptation" meme   87
3.3. "Here's How America Uses Its Land" meme   90
3.4. "Obama and Trump" meme   92
3.5. "Three Minutes Later" meme   94
3.6. "Settlers: We Have Culture" meme   96
3.7. "Settler-Colonialism" meme   98
3.8. "Indigenous Women" meme   99
3.9. "Elizabeth Warren" meme   102
3.10. "True Spirit of Thanksgiving" meme   103

Acknowledgments

I write these acknowledgments with a good heart and my deep regard and gratitude for all the mentors, friends, family, colleagues, and cohort who have walked with me on the path of writing this book. This book represents not only my work but all that has been invested in me by my communities. I hope this book in some small way reciprocates their support, teaching, and generosity. As I reflect on the many years that went into the research and writing of this book, I am especially cognizant of the political, cultural, and personal events that in conscious and unconscious ways shaped this project. I remember the day I told my mother the estimated date that the book would be published, and she told me that she would not be alive to see it in print. I brushed her off, telling her she would live many more years, but she was correct, and the devastating loss of my mother echoes throughout this text. The COVID pandemic is also present in several ways. The pandemic slowed my progress and left me feeling isolated. In this, I was reminded repeatedly that first, Indigenous knowledge-making is made in community, and second, I am a social person who needs interaction with other writers. I am grateful to my entire social media community for helping me maintain connections during this difficult time.

I am grateful to so many who made up my graduate experience at Arizona State University for their wisdom and teachings, including my dissertation committee, Keith D. Miller, Shirley K. Rose, Patricia Webb, and my graduate advisors Alice Robinson Daer, Karen Adams, James Paul Gee, Maureen Goggin, Peter Goggin, Doris Warriner, Kathleen Lamp-Fortuno, Daniel Gilfillan, Cora Fox, Joe Lockard, Elisabeth Hayes, Robert Bjork, Elenore Long, and Sheila Luna. Although this book is not written from my dissertation, the seeds of this project were planted as I developed and grew in my identity as an Indigenous scholar, teacher, and speaker at ASU, and I am thankful for the community and resources there.

I spent seven years at the University of Alabama, and I appreciate my colleagues in Composition, Rhetoric, and English Studies, Amber Buck, Amy Dayton, Luke Niiler, Alexis McGee, and Karen Gardener, who inspired me and kept me going with kind words. The Department of English and the College of Arts and Sciences provided resources for the writing of the book, including mentoring, travel grants, and the space to teach Indigenous rhetorics and methodologies. My experiences teaching graduate students at Alabama were incredibly rewarding, and as much as I mentored them, they reciprocated by helping me develop my writing in productive ways. I am humbled by the opportunities I had to learn with them.

While I am relatively new to the University of Illinois Chicago, this is where the time, space, and support to complete the project were provided to me by my new colleagues. I especially appreciate my Gender and Women's Studies chair, Jennifer Brier, for her careful feedback that helped me address two points where I was stuck, and for mentoring and encouragement in the writing process. Additionally, I am grateful to my Communication chair, Zizi Papacharissi, for ensuring I had the time and space to finish this project, as well as her expressed belief in me as a scholar and the project as valuable. The UIC Institute for Research on Race and Public Policy hosted the writing retreats that supported me in putting the finishing touches on this text.

This book would not be possible without Rachael Levay, former editor-in-chief, and Laura Furney, managing editor, at University Press of Colorado. Their support, feedback, and suggestions have been crucial to the final project. I am grateful to the editorial board for their support and approval. I will be forever grateful to the anonymous reviewers who were so generous with their feedback. I was inspired by their close attention, and this book is made so much better with their comments and recommendations. Everyone at University Press of Colorado has been responsive and helpful, and I could not have had a more positive experience working with an academic press.

I thank artist Jeffrey Veregge for his willingness to participate in the project and his contributions and generosity. I have long admired him and was thrilled when he agreed to an interview. His artistry and perseverance inspired me, and I hope you are inspired as well. We lost Jeffrey in April 2024 from complications to his long illness. My heart goes out to his family, friends, colleagues, and everyone whose lives he touched with his art and spirit.

So many people have supported me and this project in less direct but no less important ways. I want to thank my sister, Cathy, her husband, Mark, and the entire Eley clan for their cheering of milestones and hosting of holidays. You live and model generosity for us all. Cathy, especially, has been in my corner every step of my life, but particularly through graduate school, the tenure track, and manuscript writing. I may have doubted myself from time to time, but she never wavered. I thank my father, Don Edwards, for instilling in me a love of learning. He and my late mother, Helen, always believed in my writing, and I hope the completion of this project has made them proud. Academia has brought many inspiring people into my life, and I am especially grateful to Amber Buck, Frankie Condon, Michelle Bachelor Robinson, Katrin Tiidenberg, Michael Burnam Fink, John Carter McKnight, Dawn Opel, Ryan Shepherd, Brent Chappelow, Nancy Roche, Rebecca Robinson, Andrea Severson, and Abby Oakley for long conversations, warm meals, and good company. My deepest thanks to Amber Buck especially for many writing sessions over coffee or wine at my kitchen table and many excellent suggestions for this work. Without Amber, this project would not have happened. Thank you all for being in community with me. A community that was especially important to my development as an Indigenous writer and thinker is the American Indian Caucus at the 4Cs, and especially my mentors and friends there, Malea Powell, Joyce Rain Anderson, Kimberly Wieser-Weryackwe, Lisa King, and Andrea Riley Mukavetz. To my found family, Monica Boyd and Erik Bogner, someday we will make that New Orleans trip together, but in the meantime, I am so grateful for the love and belonging you have given me. I want to express my gratitude to Alyson Dizon, Alishiana Uyao, Adamaris Sanchez, Johanna Dawn, Itxhel Montaño, Catherine Ortega, Ryan Autenrieth, Abby Butler, and Sandra Coleman—my first Gender and Women's Studies students at University of Illinois Chicago. I know you will go on to change the world.

Finally, a note here to my partner, Chris Cowles, who has supported me since we were teenagers having long conversations while eating French fries on the high school quad and who has read virtually every word I have written since those days: thank you.

Hachi hullo li—I love you all.

Preface

I wrote in my acknowledgments that I am deeply grateful to the reviewers of this manuscript, who offered generous and helpful feedback. I want to draw attention here to the recommendation of one reviewer, who suggested that because I am Choctaw, I should revise the book so that it represents Choctaw rhetoric instead of the more general "Indigenous rhetoric" I had originally envisioned. After thought, I embraced that suggestion and added more elements of Choctaw culture and identity, but I want to explain here why thinking about this book as Choctaw rhetoric is complex and political. I have always known that I am Choctaw through my mother's family. When I was growing up, though, no one said, we do this or we do that because we are Choctaw. We just were, and it was not until I was older and reflecting on the writings of other Choctaw—writings of life and history—that I realized how much that culture was also a part of my lifeway and my experience. In the next chapter, I will tell the story of how that worldview and lifeway intersected with my desire to attend graduate school and become a scholar and how that caused me to rethink how I was going to perform academia. But before we get to that, I need to say a few things about why this book represents Choctaw rhetoric and how its writing is a political act of survivance.

I am not able to tell a detailed pre-contact Choctaw story, as there is not as much as I would like in the way of ancient historic record about the pre-contact Choctaw. We know some things about them, but tribes that have been removed from their land lose culture along the way. Tribes that adopt Christianity may abandon older beliefs, because they believe the Bible commands them to. Language is lost, stories change, and new stories emerge, rituals fade. And historical observers may see us differently than we see ourselves. As James Taylor Carson (1999) writes in *Searching for the Bright Path: The Mississippi Choctaws from Prehistory to Removal*, research practices have led to some conclusions about the peoples of the southeastern United States that impose the values of the researchers onto the culture. For example, we know that Choctaw women attended crops, land, children, and elders, and we know that men hunted and made war, and yet there is no evidence to suggest that there was not overlap between the roles or that women's work was valued less than men's. Still, by earlier scholars and historians, Choctaw society was seen as rigidly gender-bifurcated, with men having dominant "authority" while women had mere "influence," where the language itself undermines what was arguably a complementary society (17).

Michelene E. Pesantubbee reminds us in *Choctaw Women in a Chaotic World: The Clash of Cultures in the Colonial Southeast* that beyond that the Choctaw culture was matrilineal and women managed land and households, we know little about their other roles in culture. Also, while the Choctaw carried totems, and we have traditional tales of a Sky Woman and a Sun spirit, Hushtahli, postcontact there remain few ceremonies and rites to practice beyond the sport of stickball and traditional dances. Pesantubbee argues that (male) scholars found the Choctaw peoples and culture "unremarkable" and (male) historians largely ignored the Lower Mississippi Valley because of its early assimilation and lack of centrality in the narrative of nation-building: "The lack of interest in the unexceptional characteristics of Choctaw society and inattention to women's roles in the Lower Mississippi Valley have resulted in superficial or uninformed statements about Choctaw women up to the present time" (11). So, while today's Choctaw women hold significant roles in their families, current tribal leadership is largely male. And while small remnants of "superstitions" and Sun worship linger, the modern Choctaw are predominantly Christian. I can only share what I know through the women who raised me, and beyond that, I largely know Christianity and patriarchy. So, if writing Choctaw rhetoric means describing rhetorical performances identical to those that existed before assimilation, I cannot offer that.

I do not want to say that because the Choctaw people and their culture are not frozen in amber that I am representing a homogenous version of Indian Eurocentric imagination (Vizenor 1999). What I do want to say is that, as Carson says, the "Choctaw story is especially complicated because it involves not one culture but several—the indigenous [sic] one plus a French one, an English one, a Spanish one, an African one, and an American one" (5). So, being Choctaw today means having integrated some cultural aspects and been assimilated into others. We are a living people, and we continue to survive, persist, and create. The rhetoric here is not Choctaw because I had it all handed down to me through the generations; rather, I can make Choctaw rhetoric because I am Choctaw—it is my representation that is political, because I am surviving and making after assimilation. As a Choctaw woman, I claim and hold space to speak and to teach and to tell stories as Choctaw women, aunties and clan mothers, have done before me. I am made in the Choctaw notion of the power of women.

In this book, I share stories. I have studied my culture, and the stories I relay to you here were mostly not told to me by my family. Rather, these stories come from tribal cultural teaching efforts and published collections. And this took some detangling—some decolonial practice to arrive at these stories; because of the influence of Christianity and Christian missionary education, some of the stories have biblical influence. For example, older historians say that the pre-contact Choctaw believed in a Great Spirit and a Great Evil—this information is even present in Wikipedia. Yet today's scholars disagree and argue that while the people had a great deity, the evil spirit older scholars write about was one of many supernatural spirits in Choctaw cosmology (Akers 2013; Clayton 2020). My family are Christian Choctaw, Christianity being central to Choctaw public life since the eighteenth century, but my family carries remnants of older beliefs as well (O'Brien 2005; Tingle 2003). In this book, for example, I describe my mother's fear of owls, and I share the Choctaw story of the Owl Woman. I tell the story of foxes, the origins of corn, the rabbit's tale, the rabbit and the bear, and the little people. There are important ideas to take away from these stories: such as that "all beings and creatures of the earth [are] interrelated—brothers and sisters—and they were interdependent on each other. If one group of beings suffered a drought or famine, all the other parts of the earth would suffer too, because each was dependent on the other" (Akers 2013, 68). In addition to this emphasis on communalism and relatedness, reciprocity, generosity, hard work, humility, and family are central to Choctaw understanding of the world (Akers 2013). I hope, through this book and these stories, you will become familiar with Choctaw rhetoric and Indigenous rhetorics, as I practice them here in good ways but also as political identity work: survivance and resistance to narratives that Indians are a solely relic of the past.

INDIGENOUS VOICES IN DIGITAL SPACES

INTRODUCTION

Doing Storytelling as Epistemology

The Gift of Corn
Long ago, two Choctaw men were camping along the Alabama River when they heard a beautiful but sad sound. They followed the sound until they came upon Ohoyo Osh Chishba, Unknown Woman, standing on an earthen mound. The men asked how they could help her, and she answered, "I'm hungry." The men gave her all their food, but the lady ate only a little and thanked them with a promise.

"Tell no one you saw me. I will ask the Great Spirit to give you a gift. Return here at the new moon," she said. The Choctaw men went home and said nothing.

At the new moon, they returned to the river as instructed, but Ohoyo Osh Chishba was not there. In the place where they had seen her, though, stood a tall green plant. That plant is corn, and it is a great gift, indeed! (Nittak Hullo 2021)

I begin this chapter with a Chahta[1] story of the Unknown Woman, included in the Christmas card that was sent to members of the Choctaw Nation of Oklahoma from the tribe's chief, Gary Batton, and assistant chief, Jack Austin Jr., in December 2021. Annually, the Choctaw Nation of Oklahoma, of which I am a member,[2] distributes Christmas cards that include a Choctaw

https://doi.org/10.7330/9781646426478.c000b

story and a Christmas tree ornament that represents the story. In this book, I have included a few of the stories from the Christmas cards, some from the *The Biskinik*, which is the Choctaw Nation's newspaper, as well as stories from my mother's small archive and from other Choctaw storytellers. As I have added these stories, I have preserved the original spellings and punctuation. The stories are interesting to me because they are my heritage but also because by closely reading them, it is possible to see the influence of Christian missionaries, the integration of settler education, and the creeping of modern culture on the tellings of these stories. The Choctaw were subsumed into settler colonialism early in the assimilation and genocide processes, by treaty and by the removal. In this text, I argue that the variations in these stories are an important part of Indigenous identity—that the stories are flexible enough to be teaching stories but also memory stories and history stories, and they are at their heart identity stories. As Thomas King writes, "the truth about stories is that that's all we are" (122). I argue here that the truth about stories is that they are everything we are: history, culture, identity, kinship, faith, resilience, sovereign peoples.

The woman in the story who gifts corn to the people is an important figure in Chahta culture as well as many Native American cultures across the United States. Across Indian Country, she is also called the Corn Goddess, or the Corn Maiden, or the Corn Lady. In different tellings of Chahta stories, she is considered herself the Great Spirit or the daughter of the Great Spirit. Later in this book, there is an expanded and older version of this story. In some ways this story has been condensed to fit on a Christmas card while retaining older details like that the call of the woman was beautiful but sad and that while she was offered all the food the men had, the woman only took a little. There are, however, significant differences, as I mentioned. A rhetorical question I have is, why does the identification of her as a deity vary across the stories? As a storyteller, I vary details to highlight the significance and purpose of the story. For example, if I wanted to downplay pre-Christian Chahta beliefs, I would not mention the fact that the woman represents our Great Spirit or the daughter of our Great Spirit, because that would not be consistent with the settler-colonial patriarchy the tribe, and popular culture, embodies now.

I am using the Chahta Corn Goddess as an example that there are stories that have been told the same way by different storytellers and told differently by the same storytellers to emphasize different aspects of her. These different stories are constructed by different storytellers, art, design, cultural affinity, and kinship. Such is the story of the Chahta Corn Goddess, beginning with the two brothers. They might be impoverished, or not; or they might be sons of a chief; sometimes they're warriors, but two brothers went in search of food because they were hungry. Either in reality or sometimes in a dream, one of the brothers kills a hawk, or it could be a crow, and he roasts it. According to all the stories, it is delicious. One or both brothers are approached by a woman who is sometimes elderly but always starving. They give her food, and after she eats it, she transforms into a beautiful woman. She may have long dark hair or golden silken hair. She might wear pearls around her neck or all the way down to her feet. She rewards the brothers with a corn stalk or sometimes corn seeds.

Sometimes the brothers receive the gift right away from the hand of the woman, sometimes they are told to return in a week and find the stalk growing in the ground, sometimes it is a year of waiting. The brothers either remain hungry during this waiting time and are rewarded for their waiting, or not. Her golden hair could be the cornsilk and her pearls the corn kernels, or she could simply be a beautiful woman, or she is the daughter of the Father Sun and Mother Moon. Or not. In whatever way the story is told, it explains how the Chahta received corn, our most valuable and revered crop. A Native person can see an image of Corn Goddess and know that Corn Goddess holds those similarities, differences, nuances, and contradictions within her whole meaning—which is unlike western storytelling. I know the story of George Washington and how when he was a young boy, he cut a tree with his axe. When confronted by his father, Washington told the truth, and we are all to aspire to this virtue. That is the lesson. The legend has little variance because it has a single meaning. It is also a cultural rhetoric, reinforcing what in American thinking are great virtues, such as admitting to a lie. But Corn Goddess is a cultural knowing and a relational knowing. It is Indigenous knowing, shared, told, retold, described, historical, mythical, a teaching-learning story, and containing many more "thick" (Cottom 2019) layers of meaning about and around what is valued, what is meaningful, and what is collective identity. Corn Goddess tells us many things about our history, including the value of corn to Native Americans—a vital crop for farmers. Corn Goddess teaches gratitude, the gift of corn, the value of resilience and persistence, and more. And as I said before, Corn Goddess is not isolated to the Chahta, as corn goddess or corn maiden stories can be found across Turtle Island.[3]

But let me take a step back for a moment and talk about research, what it means to research, what methodology means, when and by whom it is applied, and for what reason. Scholars perform research to gain knowledge.

A research paradigm is the underlying beliefs and assumptions, agreed upon by scientists or researchers, as to how problems should be understood and therefore solved or addressed. As Shawn Wilson writes in *Research Is Ceremony: Indigenous Research Methods*, "as paradigms deal with beliefs and assumptions about reality, they are based upon theory and are thus intrinsically value laden" (2008, 33). Reality is subjective, and the study of what is knowledge is an interrogation of the agreed-upon reality. Once while in the countryside, my friend Daniel saw a fox. He posted to his Facebook friends the question of what it meant to see the fox cross his path. Daniel is not Indigenous, and neither are most of his friends. His friends responded cleverly, and some with quips about what the fox said, a cultural reference to "The Fox (What Does the Fox Say)," a Norwegian pop-electronica song from 2013, the video for which has been seen upwards of a billion times on YouTube. What the fox says is western cultural currency—a popular song. For the Chahta, the fox is associated with shadows and can be a creature who moves between shadows and worlds. In scientific classification, the fox is in the Canidae family. In biological study, foxes do not speak in human language. For Chahta, foxes may speak. What is the agreed-upon reality of the fox, then? This is an ontological question: what is the real fox? In science, the classification is a value-laden system, as to classify is to define, catalog, quantify, and stabilize knowledge. It is the colonization of creatures to settle reality. For the Chahta, Fox has thicker, perhaps the thickest of meanings; Fox delves into the dark things that Chahta know about life, afterlife, souls, and the nature of good and bad. Fox is both frightening and reassuring that there is a deep and wide life—it is challenging and is not comfortably contained in one idea—Fox strains at the boundaries of meaning. Fox is definitely not settled, and it does not matter to the Shadows if Fox is classified by westerners as canine. I bring up the fox story as a storyteller here in discussing the purpose of research as to explain that there is a quantifiable, definable, catalogable, settled reality that is effectively described by western knowledge production. But this western fox is not the fixed reality we agree upon, because in Chahta cultural knowledge production, Canidae is not Fox. There is not one reality nor one way to know Fox, but if I am writing in an Indigenous research paradigm, I am writing about the cohesive tension of stories and meanings that is Fox.

I want to say here that this book is written from an Indigenous worldview of good relations. "Good relations" is the practice of coming to a collaboration with openness, good faith, sincerity, reciprocity, and a respect for others as relatives with each other, the land, and its ecologies. I write here in a good-faith effort, with respect for and openness with all my relations, including the readers of this text. I commit to telling the stories here carefully and with concern for the representation of Indigenous peoples found within. I also write with equal care and respect for those mentors and colleagues whose advice and experiences I discuss that have shaped how I research, write, and think. In this book, I instantiate this worldview as I tell these stories, pairing my methodologies with explicative text on how the methods and analysis do research work as I write about them. I theorize and apply Indigenous methodologies and epistemologies to case studies and reflect on what can be learned about ourselves and our world through these practices. In this book, I offer four case studies describing and doing Indigenous and digital rhetorics. This book is conceived as co-construction of knowledge between these projects, their participants, these words, and the reader. I invite the reader to walk with me as I explore both what it means to experience being Indigenous in digital spaces and the possibilities that open by applying Indigenous methods and practices to non-Indigenous contexts.

First let me tell the story of how I arrived at my Indigenous scholarly identity and began doing Indigenous-oriented digital rhetorical work. I am telling this story to situate myself as an Indigenous woman, daughter, technologist, and rhetorician. Here, I explain and describe my journey, identifying myself as a professional technologist, and then as an academic, and then as an Indigenous academic. My hope here is to do several things. First, I want to discuss and demonstrate how it is possible to shift thinking from a western worldview to an Indigenous one, and to shift from a western worldview to an Indigenous lifeworld. My recounting of how I arrived here is not about bringing Indigenous peoples and their cultures into the digital present, for we are already here and not relegated to an analog past, or worse, the distant past. As Elena Ortiz of *The Red Nation* podcast reminds us, even relegated as our peoples are to wings of natural history museums, we are in fact not dinosaurs (Ortiz 2023). No, my recounting is about how to break away from western ways of knowing by doing self-reflective and self-interrogative identity work within digital research frameworks. I want to talk a bit about our teaching and learning and how western practices are replicated in students, at times writing over valuable, existing cultural ways of being. I want to demonstrate doing cultural rhetorics, as I situate how I do knowings and meanings within my own culture(s), digital, Indigenous, feminist, and otherwise. Finally, I want to demonstrate writing research as story, building on the work of other Indigenous scholars and the kinds of thick and robust meanings Indigenous methods have to offer.

I did not set out to write my first monograph about indigeneity, or being Indigenous, online. In fact, my first several scholarly projects, including my dissertation, examined gender online, through a feminist lens trained on my software industry experience. As a graduate student and in my first few years as junior faculty, I thought my first book would be a feminist theory text about gender and technology. For me, with more than a decade spent in the technology industry, both writing code and managing software development projects, I thought what I had to offer to research conversations was my uniquely gendered experience with the technology industry and its master narratives encoded in its design processes. I saw "project manager" as my identity because I had been making computer programs since childhood, beginning with BASIC and then following with spreadsheet macros and finally database design. You can get a lot done with a little bit of knowledge about how Microsoft products work. You can get even more done with some platform-specific classes, which my employer at the time paid for. "Project manager" is how I knew myself for a long time. This way of knowing myself was largely settler-colonial, and here I will explain how. As a child, I was the daughter of a Chahta woman,[4] but that had little to no meaning in a settler-capitalist world, other than its cultural associations with alcohol, poverty, casinos, and violence, which are stereotypical and racist settler-colonial narratives of Native American lives. But in general, in the capitalist paradigm, children are not fully participating members of the settler-capitalist society until they finish their educations and assume an employed role in that same society. Children grow to become lawyers, doctors, truck drivers, software developers, teachers, and service workers. You do not *become* in settler capitalism until you are a wage earner and product consumer. I started working in technology at sixteen and *became* a software development project manager as an adult.

As a technologist, I have been a part of both public- and private-sector development teams and implementation projects. I have worked with large numbers of coworkers in lumbering enterprise environments, and I have pulled my weight on a small start-up team. Although it has been many years since I did this kind of work to earn money, I can still play with databases, I can write some code, I can design and administrate websites, and I can maybe help you fix your email. I am a technologist, or so, as I said, my settler-colonial, capitalist worldview I had been programmed into tells me. Then, after my time in the technology sector, I enrolled in graduate school, and my identity began to change. Over time, my perception shifted as feminist research methods and feminist theory courses had me thinking more deeply about myself and reflecting on my role as a researcher. Maybe having some knowledge about technology was not the only thing I was bringing to the field, or maybe not even the most important thing. What I brought, upon much feminist reflection, was myself, and myself was many things. Yes, I identified by my job, as most people simmering in settler-colonial economies do. But graduate school, with its practice of self-identification and feminist self-reflection, helped me disentangle myself from the neoliberalization of identity and discover myself as a complex, thinking, learning, growing person. Graduate school helped me recover from being only a capitalist production entity and reminded me that I was also a Native person, whose heritage had never been surrendered to the settler-colonial machine.

One of these emergent discoveries was my realization that my Chahta cultural identity and cultural practices greatly informed how I thought, communicated, and made and negotiated meaning(s) and knowing(s). In the past, I may have identified myself by my role in the economy, because that is how we identify people in the West ("And what do you do?" being the relevant conversational question), but once I started demonstrating thought processes and collaborative processes in a graduate cohort, my Indianness[5] emerged. You can meet me in person and see a person of mixed heritage, but once we start collaborating, I cannot, even if I wanted to, hide my Indigenous ways of being in the world, because they are how I make knowledge. Through my courses in rhetoric and composition, I came to understand the importance of situatedness, space and place. And I began to see, when working so closely and deeply with my colleagues and professors, that many things about how I think and how I speak are different. My Native-Americanness was no longer relegated to my personal and family life; it was drawn into the forefront, because it is long established in the humanities that we negotiate knowledge practices and knowing through our many and varied positionalities. The field of cultural rhetorics, specifically, speaks to our constellated cultural constructions, through which we know, identify, and make meaning (Powell et al. 2014). We are our stories, particularly in Indian Country, where stories hold together in tellings, retellings, and sometimes contradictory tellings, in a kind of communal knowing, which challenges the "rugged individualism" of white, western realism and neoliberalism. I wanted, then, to unlink my job from my identity and return to the collective identity, a layered and complex identity, by which I always knew my private self. In other words, I wanted to reconfigure and unify myself, not as a consumer-worker in the technology industry but as a relative to others, for whom value comes not from industry but from living and learning in my Indigenous lifeways, as well as the natural world.

I offer here a simple example of remaking myself by discussing here a bit about confronting my name. During my doctoral program, I realized I heard my last name from students many times a day, when they called me "Professor Cowles" or "Ms. Cowles." At the time, I was using my husband's last name—a choice I made in my early twenties to defer to the traditions of my in-laws' patriarchal family and practice. But hearing his family name when I was being addressed every day by dozens of people did not seem to align with my self-identification as a Native teacher, learner, and community member. It did not reflect my own matrilineal family structure either. Up to that point, I had been in industry all my adult life, and my name was not something I thought about. I was always known by my first name or my user ID or handle or email address, or variations of both. A user handle is its own kind of identity; it is who you are within a delimited context of a system. It is created for the system, along with those of all the other members of the system (or a network), who have their own system-specific identities. It ties your name together with your competencies, your coworkers, and your specific space, place, and time. When I am working, and I am being called by my user handle, I know what is expected of me, I know I am one person in a specific network, and I know I am assessed for my abilities and contributions in the context of that network. This is another form of self-identity—myself as part of a larger group—yet only within the context of the network am I known. In this context, nothing else about me matters, especially not the personal. This applies in many ways to everyone, whether they come through industry or not. We have, as we joined networked society, adapted to a digital ID and a networked identity, across the platforms we use, like Twitter and Instagram, but also Blackboard and other learning management systems (LMSs), and even our e-commerce practices. But as Galloway and Thacker write about Geert Lovink's work, "informatic networks are important, but at the end of the day, sovereign powers matter more" (2007, 1). After reading these scholars in coursework and considering my own position, I was beginning to see myself beyond networked identities, my digital life, and my keyboard.

My graduate-school (re)emerging identity of being a Native teacher, scholar, thinker, and knower raised personal feelings for me about my Indigenous identity, which had never been reflected in my industry jobs. I had occasionally referred to my indigeneity in passing, mostly in terms of why I could not work extra hours on a particular weekend. Here, I offer a side story to understand the rare moments in industry when my indigeneity became visible. Even though I worked many hours a day, most days of the week, at the start, I occasionally had family obligations, like cooking for a party for my sister, Cathy. I once told an executive that I could not work on a weekend because it was my sister's birthday and I was cooking for the family. He said he would hire a caterer for me so I could still work the weekend and just take a break to pop in and enjoy the party, and then come back to work. In one view (if you squint, maybe) this offer is kindness, and from another view it is an appalling instance of neoliberalism. But I am Indigenous, and expressions of commitment and connection to family cannot be farmed out to a third party, no matter the intention. The point was not that there was food for the family dinner, the point was that I made and provided the food with my own hands and skills, food that was taught to me by my mother, handed down from her extended family network. So, a few times, my personal identity clashed with my industry identity. And my Indigenous identity was read as inconvenient, and I was viewed as too "stubborn" to set aside culture for the good of the project or the company. I was not seen in industry as a cohesive person, because I was a worker in a capitalist, technolibertarian context where culture is devalued and even rejected as an impediment to globalism.

Now that I was becoming a fully constelled Indigenous person in graduate school, I found myself questioning my identity and its connection to my work. As I have described here, I have always thought of myself in relation to my technology career. The technology industry and its culture, a toxic mix of neoliberalism, technolibertarianism, heteropatriarchy, and capitalism, reinforced my identity as tech worker. By contrast, my identity of Native teacher-student-thinker-learner is not an Indigenous career; rather, it is a significant role situated within family and culture. Knowledge is not the books and empirical processes of a scholar, or the stored data bytes of a technologist; it is an amassed knowing, a constellation of collected experience through the sharing of stories, the details of the lives of the community, and a kind of collaborative building (Powell et al. 2014). Indigenous knowing is taught and constructed across generations in cooperation with the land, the community, and the extended family. Here, as a graduate student, I know myself as Helen's daughter, Esther Belle's granddaughter, a sister to Cathy, and a cousin to many. I am a Native teacher-student-thinker-learner, like the people in my extended network before me.

As I write this nonlinear narrative, I thank my mentors who introduced me to feminist theory and feminist scholarly practice. I thank those who taught me network theory, kinship knowings, and postmodern theory. I thank my mother, who taught me how to garden, sew, and refinish and repair old things; how to cook together as carework; how and why to tell stories; and how to be in the network of our extended family. When I grew into my role of Native teacher-student-thinker-learner, I did so by working through feminist metacognition and feminist self-reflexivity. My own cultural practices of relating to others and situating myself within the context of my identity and my work, given the western framework of single-scholar knowledge-making, caused me both joy and conflict. I wanted to be a scholar, but I also wanted to be myself. This moment reminded me of the lunch party celebrating my completion of my master's degree, where my mother gave a brief speech. She said she wished me all the success in my plans to enter a doctoral program, and then she reminded me never to forget my family or what they had taught me. I often thought back to that moment while completing my doctoral coursework, especially when I was feeling isolated. It was in this period that I resolved the issue of my name, with support and encouragement from Dr. Karen Adams, with whom I studied sociolinguistics, and with a consultation with my mother. Dr. Adams reminded me that names matter, and we talked about matrilineal names and their power. I asked Mom what she thought about using Tekobbe as my name, and she said she thought it would "be an honor to remember the family." Needing her permission, or at least her approval, is an example of how I do not make major decisions for myself without considering their impact on my family and community. Mom loved the idea, so I went to court and changed my last name from my husband's last name to my family's name, Tekobbe, a name that dates to before the Dawes Rolls.[6] My thinking was that if I were to be a Native teacher-student-thinker-learner-grower, I would be so as Indigenous, making visible my own voice. I would foreground my Indigenous identity and therefore my Indigenous ways of being and knowing, even as I continued thinking about and writing about the digital and the social. This is where the slow shift in my scholarship began, in a moment where my identity tacked in from multiple waypoints, from technologist, scholar, Indigenous woman, and teacher, and I built myself an authentic place to stand (Royster and Kirsch 2012).

This self-identification that prioritizes my indigeneity is important, because while I emerge as a more authentic version of myself and my ways of knowing, I find myself negotiating race in new and challenging ways, in terms of phenotype, of discursive practices, of collaborative practices, and of meanings and definitions of family, to name a few. Here is a fact about myself: I am Cindy Tekobbe, and I have blood quantum "evidence" that I am a member of the Choctaw Nation of Oklahoma. Blood quantum is a tangle in itself, because blood quantum is one of the genocidal processes inflicted on Indian Country by the settler state. Blood quantum is a colonial technology designed to mark off and then set apart Indian folks onto federal reservations. While all historical efforts to define people by drops of blood are racist, of the racialized identities in our nation, Native American is the only one quantified by both tribal rules and federal law and policies relating to who one's parents and grandparents are—by "blood."

I often present this citizenship-membership status when I introduce myself as an Indigenous scholar—not the colonial blood quantum, only that I am a member of a sovereign Native nation, the Choctaw Nation. When I meet other (white) scholars, generally their first response is to assess my appearance to determine my Indianness. Often, I am told that I do not look very Indian, that they "never would have known" without my making a point of it. Or sometimes when I self-identify, they look at my face and ponder that there is a shadow of Indianness in the flare of my nose or the prominence of my cheekbones. These are supposedly concrete phenotypical markers that I must meet in order to be accepted as Indian (Arola 2017). What does it mean to "look" Indian, anyway? What does that look like? As Michele Leonard (2023) writes in "You Don't Look Indian" from the *Unpapered* collection edited by Diane Glancy and Linda Rodriguez, Hollywood has for generations perpetuated ethnic stereotypes about Indianness, and those are the most familiar faces to us, the audience. Recently, *Reservation Dogs* (2021–2023), a program about Native Americans that has the participation of Indigenous people in its writing and production, has challenged the Hollywood stereotypes about what Indians are like (Leonard 2023, 126). Still, stereotypes persist, and I am sensitive to the perceptions of others that I am less (or not at all) Indian because of my appearance.

This whiteness-centric assessment of Indianness I will discuss in more detail in later chapters, but the takeaway here is that when I escaped one narrow box in my identification journey, I found myself in another. This time, a colonial, legislated box. Watanabe writes about the colonial and Indigenous tensions I am describing here, in "Critical Storying: Power through Survivance and Rhetorical Sovereignty" (2014). Our stories are not to be used to essentialize us through deficit narratives about poverty and underperforming students and blood quantum that is reduced with each subsequent generation until none of the Indian is left. This is why it is important to center Critical Indigenous Research Methodologies (CRIM), because the question of identity is also a question of sovereignty (Brayboy 2005). Sovereignty is power, the power to define ourselves through our stories, and our stories are our theoretical practices and learning practices. It is through these stories that I know who I am, not by the flare of my nose or the color of my skin.

As I have explained, Native American identity is legislated, defined in various ways by law, both of the settler state and tribal governance. But we are our stories and our communities. A Native person when meeting another Native scholar will often ask for tribal affiliations and family names rather than discuss eye color or skin color. We do this to locate each other in the community of peoples and to determine if we share or overlap in those communities. Identity is not, then, what you look like but who your people are. An example of this can be read in the exchange in chapter 4 in my interview with Indigenous artist Jeffrey Veregge, when we exchange community locations and tribal affiliations. This practice is sustained by traditional collaborative identity work: we are who our people are, we are where we come from, and you belong to the community that claims you. Affiliate identity and family identity carry the tension between anti-colonialism and colonialism, where blood quantum is claimed as a valid identity practice in retaining tribal identities and is also legislated by the settler state in its genocidal practices of counting and controlling Native populations.

I am an enrolled Choctaw through my mother, who was also an enrolled Choctaw. My mother demonstrated this kinship when she enrolled my sister and me in her tribe with her genealogical records. Therefore, I have tribal identification to demonstrate my kinship, credentials that make me Choctaw for both my tribe and my university employer. When we fill out our equal employment paperwork, we, unlike other groups identified on the applications, have tangible "proof" that we are who we claim to be, if only for the settler state.[7] I cannot count the number of times I have been told that I do not "look" Indian, as if (1) one's Indianness can be determined by common appearance and (2) as if white folx are arbitrators of race, culture, and community. Why is my skin lighter than they expect? Why are my eyes blue-green? I say that I have been told that I am also of Irish, English, and Dutch ancestry, and then I become, by the assessment of some, not Indian enough. As my friend and colleague Amber Buck reminds me when I complain to her that I am caught in this racial-political tangle, "this is how whiteness works." Whiteness is normative, whiteness is the default, whiteness subsumes and erases other identities—why else would I "claim" Native heritage, claim being the operative word, given that whiteness decides whether my claim is authentic, when

I clearly can also be white? On the other hand, the fact that I am of an identity other than white, one that has been subjected to centuries-long efforts of the settler state to remove, absorb, erase, and eliminate, also makes me inferior, because this is also how whiteness works. If you could be white (or presenting as white), why would you not?

For me, I reject the divisive and genocidal blood quantum. Instead, I know I am Chahta, because my mother was. My grandmother was. My aunt and uncles were, and their children are. I am of my family, again, of the people who claim me, and I speak, research, and write as an Indigenous woman not because of federal efforts to catalog and control Indians but because of these family relationships. In fact, these days, I often decline to answer questions about my Indianness when someone is speaking about how I look or other factors, like my education, that do not align with their notions of American Indian. I do not like to get into discussions with white-identifying folx of whether, with my mixed ancestry, I am Indian enough (in these types of discussions, where someone else assumes the right to define me, can I ever be enough?).

Another point of difference for Native scholars is that one's own community might censor a Native person for speaking a narrative in a way that is subsumed in, as Powell and coauthors write, a "prime" narrative (2014). In other words, because identities are raced, and racism stereotypes and condenses these identities into one nominal representative, sometimes when I speak for myself, what I say becomes what every Indian says and thinks. I feel this responsibility and risk acutely when I write, speak, or teach. Some of this feeling of precarity, of fear of my words carrying too much significance, for certain has to do with being raised to not draw attention to my Indianness, from parents who grew up under Jim Crow and my grandmother, who experienced various forms of racism and loss related to things like land and resource allotment and personal autonomy. I demonstrate some of this racial complexity and prime narrative in chapter 5 when I offer the case study of Payu Harris and his efforts to bring the first cryptocurrency to Indian Country. I will discuss more about Harris a little further down, but what is important here is that Harris is a self-identified Lakota, but this identity eventually led to the downfall of his cryptocurrency project, because he was at one time too Indigenous, not Indigenous enough, too white, not white enough, too outspoken, and not well-spoken enough. Knowing who we are is never enough for the settler-colonial authentication system, and Harris's case demonstrates the perils of this prime narrative of white supremacy.

I also want to say that more recently, as I was revising and preparing this book for publication, there has been an effort in academia and popular culture to root out "pretendians," people who claim Indigenous ancestry but are not from Native communities—pretend Indians. Several prominent scholars have resigned or been removed from their positions in the United States and Canada. Sacheen Littlefeather, of notoriety from when she declined the Oscar for Marlon Brando in 1975, has been accused by researchers and journalists of being a pretendian. Their research on her genealogy, they argue, demonstrates that she is not Apache as she claimed. I will touch more on pretendians in my chapter on memes and my discussion of Elizabeth Warren, but I want to tread carefully. It is wrong for people who are not Native to assume positions, titles, scholarships, or other resources in academia that are intended for Native Americans. And I agree that many people may have family stories about Cherokee great-great-grandparents that cannot be verified or may not be true. I am deeply concerned about the efforts to disprove Native identity by journalists and some activists, though. I am concerned that these efforts disenfranchise many people who have Native ancestry and are engaged in Native activism and community carework but do not have blood quantum documentation or are not from federally recognized tribes. The search for pretendians should not, in my view, reinforce colonial expectations of our people. Earlier I mentioned the excellent collection of essays *Unpapered*, edited by Glancy and Rodriguez. This collection has a lot to say about what it means to contribute to and be a part of Native communities without the enrollment paperwork.

Returning to my story narrative of how this book came to be: my Indigenous identity, my personal life, and now my teaching and research lives are more authentic to me and how I see and know myself. And I was slowly making change as a scholar. In the earliest years of my studies and career, I wrote a few articles and chapters and I gave a number of conference talks on gender and technology. Then, in 2013, a friend and collaborator, John Carter McKnight, who was at the University of Leicester in the United Kingdom, found an international news blurb about an emerging cryptocurrency—a Bitcoin variant called MazaCoin—that was being implemented by an Indigenous tribe in North America, the implementation effort led by a Lakota man named Payu Harris. After following the press coverage and studying the way the technology industry media in particular were covering the story, John and I knew we were seeing something different about the reporting. It was less about the logistics and process of implementing the new cryptocurrency and more a story about the scrappy entrepreneur and his battle against the mighty federal government. Harris was a lone Indian rising from the past to save the future with technology, except carried along on the story was the neoliberal and technolibertarian culture of the technology sector. In other words, it was very much a colonial retelling. McKnight and I were not the first to discover that news media and mass media objectify and appropriate Indigenous histories and identities, but we were seeing it in the technocratic media we followed and studied, and that was new. McKnight suggested we submit a conference proposal to the Association of Internet Researchers, focused on MazaCoin, the Indigenous cryptocurrency an entrepreneur and tribal member was trying to launch on the Pine Ridge reservation.

This was the first time I had thought to bring my Indigenous perspective to scholarship. It was also the first time I considered that there might be a distinctive way of being or being-made-to-be Indigenous online. Through my examination of the buzz surrounding the rise and ultimate fall of MazaCoin, and the role Payu Harris may have had in this situation, I came to understand that digital coverage and reporting of Indigenous people in examples of online journalism has flattened Indigenous identity into gross settler-colonial and neoliberal stereotypes. I began to think about not just identity but decolonization, survivance, capitalism, neoliberalism, colonialism, and other ideologies that shape the Indigenous experience in personal, cultural, and political lives. I began to conceive of digital research from an Indigenous positionality where I could untangle the settler-colonial assumptions and tease out the complex and thick knowledge-making in Indigenous contexts. I would come to research in the spirit of good relations, I would write and speak my findings, not with western skepticism and hostility (a knowing is false until proven true), and as my people do, I would treat my work as ceremony, a system of honoring, respecting, and contributing to Indigenous community.

And I arrived here at my first research question: **How do Indigenous peoples construct themselves in digital spaces and places, as opposed to how digital medias construct them?**

And a second question followed immediately after: **Can I find other examples of where the stereotypical descriptions of the race of Indigenous peoples are complicated or subsumed in digital spaces?**

Indigenous Research Methods and Practices

These transformative experiences I describe are how I arrived at the writing of this book that applies Indigenous frameworks and epistemologies to online cultural movements across four case studies. With the findings of these studies, I contend that Indigenous peoples employ social media and digital technologies to construct their identities as modern, engaged, and living peoples rather than allow themselves to be relegated to history. I also contend that these methods can be applied to additional cases of online research in order to break western paradigms of oppositional critique and participant objectification. I argue that, as I discussed in the opening of this chapter, Indianness is persistently assessed, and legitimization of that identity is a determination claimed by white audiences, both because this is how whiteness works and because there are settler-state laws and histories that make this possible.

I argue that western thought and western theory are too narrow, and too focused on individualism specifically, to investigate Indigenous identity construction. And the decolonial and survivance practices can be applied to uncover richer and more complex interpretations of Indigenous digital practices. With this book, I seek interventions into research problems created by mainstream critique and western theory, research practices that not only flatten meaning but reify singular authorship instead of valuing collaborative texts. These conventional approaches and frameworks tend to objectify research participants, co-opt participant experiences, and, with their insistence on an oppositional framework, undermine the research process by introducing skepticism and positioning the researcher, rather than the participant, as the arbitrator of truth. In chapter 3, which is about internet memes as collaborative identity construction, I build on the work of other scholars interrogating memes as identity-building in white-centric digital spaces, and I explore how the storytelling as identity practices bring both wider interpretations and thicker relationship ties.

This book offers Indigenous methodologies and new epistemological frames to explore digital communities and technologies. These approaches are designed to help solve the problems of conventional western critique and oppositional positioning by adopting storytelling as methodology, centering good relations and relationality between researcher and participant, and ethically positioning the participants' experiences as the measure of truth. In storytelling and stories, as Linda Tuwihai Smith writes, "each individual story is powerful. But the point about the stories is not that they simply tell a story, or tell a story simply. These new stories contribute to a collective story in which every Indigenous person has a place. For many Indigenous writers, stories are ways of passing down the beliefs and values of a culture in the hope that the new generations will treasure them and pass the story down further. The story and the storytelling both serve to connect the past with the future, one generation with the other, the land with the people and the people with the story" (145–46). In other words, stories are both fixed and fluid, adapting with details added by different storytellers, or different tellings by a storyteller.

With this text, I argue that Native Americans' use of social media and digital platforms uniquely constructs Indigenous identities as living, producing, culture-making peoples, working against the commonplace narrative that Indigenous North Americans either live in isolation from everyone else or are simply a people resigned to the long-forgotten past. I argue that common forms of digital analytical research methodologies, for example, visual analysis, discourse analysis, quantitative coding, and so on, add to the flattening and confining of thick stories and narrow findings to single interpretations of collaboratively arrived-upon meanings. These thick stories and thick meanings I derive from the above-described practices of collaboratively told and retold stories, meanings, and identities. Within the layers of these artifacts are bound individual, generational, new, and old contributions to the thick meanings.

This book contributes to the field by injecting these frameworks and methodologies into digital rhetoric, which, as a field, is seeing researchers take up approaches from critical race and gender theory. This injection also impacts research ethics by expanding on the reflection and relationality from a perspective informed by feminist research ethics. Indigenous research methodology emphasizes the role of ceremony in both the daily practices of Native peoples as well as the research practices in Native communities and contexts. Ceremony, simply, is the practice and process of honoring the sacred. This notion of ceremony is tied to another Indigenous notion of good relations, meaning that knowledge is approached as created within the context of relationships between people, and those relations are grounded in trust and open-mindedness. This framing of good relations is largely unique to Indigenous research, but it is an ethical and holistic approach that would be of broad interest to cultural and digital researchers. The value of Indigenous research methodologies is that they are relational, subjective, personal, and emotional or intuitive, which has the potential to respond to the overarching concern that our research is so grounded in western notions of knowledge that we inadvertently reinscribe hegemonic structures over our research participants and their experiences. Social media is, above all things, social, and most of our digital methods do not have a way to account for the social (emotional, intuitive, personal) aspects of digital artifacts. Indigenous approaches are one possible antidote to this problem.

Indigeneity as Public Discourse

After I had begun the MazaCoin project in chapter 5, I did not have to look very far or very hard to find more journalism reporting Indigenous news using colonial constructions of Native peoples and their issues. In fact, the more I looked, the more I realized how deeply widespread beliefs and negative tropes about Indigenous peoples informed the political as well as societal landscapes. I also had thoughts and questions about the "neutrality" of digital journalism, if it was built upon settler-colonial knowledge. For example, in stories about Indigenous peoples and their political and cultural presence, I expected to see reporting of racism where it was obvious, but instead, I found that stories referencing Indigenous peoples and their issues were written with a kind of false neutrality. Rather than sharing a long and complex discussion about journalism and constructs of neutrality, I will simply say this: we live in a white supremacist, settler-colonial nation-state. We are a heteropatriarchy. Therefore, as Arvin, Tuck, and Morrill, and other scholars, have reminded us, within this construct, everyone is racialized and gendered (2013). I would add that the settler-colonial nation-state runs on settler-colonial capitalism, where production is required for membership. There is no neutrality within this matrix, only the gloss thereof.

One obvious example of this practice I found in the reporting, beginning in 2016, of Trump's racist attacks on Elizabeth Warren, who has asserted that her family stories describe a Cherokee ancestor, therefore making her a Cherokee descendent (Fonseca 2020). The issue of Warren's Indigenous identity was a talking point in her 2012 campaign against Senator Scott Brown in the senatorial race in Massachusetts.[8] It was widely reported and fact-checked, and in the end was settled with Warren apologizing and stating her Indigenous identity was based in family lore, and many experts agreeing that there is no historic evidence to demonstrate that Warren is of Cherokee descent. Experts also note that matters of Indigenous descent can be difficult to prove, with many Indigenous peoples not being included in the original rolls for a wide range of reasons (Lee 2016).[9] Still, it was determined that Warren was not Cherokee, and her case stands as an example of the tensions around claiming Indigenous identity without connecting with communities and holding relationships with Indigenous people. Donald Trump resurrected this controversy in November 2017 when, in a meeting and award ceremony with surviving World War II Navajo code talkers, he referred to Warren as "Pocahontas" (Haltiwanger 2017).[10] Yes, Mr. Trump made a negative crack about Pocahontas while honoring Native men. The racism and sexism of that moment are far more complex than just surface offense, especially given the significance of code talkers in the Native communities they were members of. These people are highly revered as warriors, culture keepers, and role models to Native peoples. Their contributions of sacred language skills are incredibly important in Indian Country. They are also American heroes whose use of Native languages helped turn the tide of World War II. Not isolated to simply offending Native war heroes, Trump continued this practice of calling Warren "Pocahontas" through his single term as president, as well as while on the 2020 campaign trail. (I write about this in chapter 4 as well.)

In the middle of Trump's speech thanking the code talkers for their invaluable contributions to the United States during World War II, Trump pointed out that the Navajo code talkers were "real" Indians and Warren was a "fake." Later, the Navajo Nation, when asked for a comment, was reported to say that they wished to not be involved in the president's conflict with Warren over her questioned indigeneity. Trump continued referring to Warren as "Pocahontas" while she was on the campaign trail for the 2020 US Democratic presidential primary election. As I revise this chapter, days before the 2020 presidential election with Joe Biden and Kamala Harris on the ballot for Democrats, Trump is still referring to Warren as "Pocahontas."

Ali Nahdee, feminist Indigenous critic, says of Indigenous female representations that Indigenous women deserve to have mass media representations they can admire and look up to, as opposed to what we have now (2020b). She argues that Disney's Pocahontas is another sexualized and airbrushed version of a woman whose history was complex and deserves more respect than she has been given in media treatment. She speaks of her "Aila test," which looks for Indigenous female characters in films who (1) are Indigenous and a main character, (2) do not fall in love with a white man, and (3) are not raped or murdered as part of the plot. She points out that with respect to expectations of media representations, it is important to bear in mind that many are complicated: perhaps the female main characters are Indigenous, but they might also be troubled characters with darker histories. That does not make their existence unimportant to Indigenous representations. She claims, "We don't have to be perfect, but we don't need to be killed all the time" (Nahdee 2020a). Sexualized and victimized representations of Indigenous women in media are complicated by the real-life vulnerabilities of Indigenous women. The movement for Missing and Murdered Indigenous Women, #MMIW, has worked to bring attention to violence against Indigenous women and the legislation that has followed (Whitebear 2020).

Trump, in his many racist attacks on Warren, makes it clear that he is referring to the 1995 Disney version of Pocahontas, the "good" version. However, in all the reporting surrounding Trump's tweets and speeches where he calls Warren "Pocahontas," there is very little written about why these attacks are racist. It is almost that because Warren has been identified as a "pretendian," there is no other wrong here. So, here I argue that these attacks are racist, because they identify Warren with a highly westernized and sexualized version of a historic Native woman. In the digital media stories about Trump's offensive nickname for Warren, I found that if racism is mentioned at all by journalists in those articles, it is couched in terms that only a few people find racist, or that only some find controversial. Here is another case where we find that media neutrality that distorts Indigenous people's struggles and slaughter and colonizes the Indigenous history of the United States by equivocating about Indigenous representations and struggles.

To call Warren "Pocahontas" is to call her an airbrushed Indian. It mocks both Indigenous history and Warren, making both mere caricatures in Trump's nationalist bluster. In particular, Trump's reference to "the bad version" of Pocahontas speaks to his self-awareness that he is pointing to a caricature rather than a historic figure whose history is bound up in the colonization and Christianization of Indigenous peoples. And the crux of it is an extension of his nationalism, his ongoing campaign to "make" America in his own image, mocking marginalized identities generally and Indigenous identities specifically. There are more examples of Trump's attacks on Native people, like his disrespecting Natives by declaring November's Native American history month "National American History and Founders Month," but again, they are beyond the scope of my claim here, even as they are worthy of attention (Armus 2019).

Turning to another recent event, during the run-up to the United States November 2018 midterm elections, a news story broke about voter suppression of Native Americans in North Dakota. There is a long history of voter suppression, which I discuss in some detail later in this book, but to provide a brief overview, the North Dakota law was one of many voter ID laws passing through state legislation that suppressed the Native vote. The US Supreme Court had ruled in favor of North Dakota to uphold a voter suppression law requiring the possession of state-issued identification that includes a physical address, for citizens to be able to vote (Hayoun 2018). On North Dakota reservations, like many rural reservations across the United States, many residents are given only PO boxes, to simplify the mail delivery process for the federal government. In other words, the federal government allotted remote lands to people, lands that were too remote to easily access, so PO boxes were implemented for the convenience of federal services. Those same conveniences are suppressing Indigenous voters in North Dakota. This is institutional racism, but these issues have largely gone unreported.

In addition to not having physical addresses, some reservation residents might not have state-issued identification either and instead use only tribal identification cards that do not list a person's physical address. This is an issue of Indigenous sovereignty in the United States. Indigenous peoples are, by treaty, sovereign in their own lands, and their identification represents who they are as Native community. Yet, these identification cards are not "official" enough to satisfy the identification changes. This is a rejection to Indigenous sovereignty, which again was not widely covered or discussed.

In addition to the potential lack of physical addresses on ID cards, Native Americans are also overrepresented in unhoused populations, where they are also unlikely to have physical addresses (Domonoske 2018). Together, these factors (and others) disproportionately affect Indigenous populations and communities, and work together to actively suppress the Native American vote and political engagement. This context of suppression gives rise to questions of the use of social media as a nontraditional way of accessing political power, as marginalized identities' and groups' means of social support, and as the self-presentation of Native American groups online. I expected to see this racism called out and discussed, and largely, it was not. There are several well-known examples of Indigenous digital activism, such as the Idle No More movement emerging in 2012; the #NoDAPL campaign resisting the construction of the Dakota Access Pipeline, which crested in 2016; and the abovementioned work to raise awareness about MMIWC (missing and murdered Indigenous women), which notably took shape around 2016.

Thinking Further

Ultimately, this book is about being Indigenous and being digital, being Indigenous and being in media, and the ways colonialism, racism, white supremacy, and patriarchy complicate building an Indigenous identity in digital spaces and on digital platforms. But it is not just that Indigenous peoples are pressed in on all sides by multiple flavors of imperialism and supremacy, it is also about new ways for Indigenous peoples to use digital spaces and tools to actively speak their truths and be present in everyday digital interactions. It is about how Indigenous peoples have been relegated to the past by colonialism, and by accessibility issues, as well as how Indigenous people are building their new digital presence. I theorize the thickness of Indigenous meaning-making to complicate the simple and one-dimensional stories of Indigenous peoples today. To get at these points, I use Indigenous rhetorics, intersectional feminisms, digital rhetorics, decolonial theory and practice, and other bodies of theory to excavate indigeneity from its oppressions.

I will also say what this book is for me. It is the culmination of work thinking about the world in networks and thinking about the world in all its relations. I make a note here about language that I think is important for a reading audience. I wrote this book thinking about ways to write myself, a digital Indian. I call myself Indigenous and Native American. Informally, I call myself an Indian, because my mother, auntie, and grandmother always had. I use *Tribal*, *NDN*, *rez*, and other Native words and phrases because my family always has. In this text, I use these labels interchangeably, and the words periodically. I would appreciate it if, when you cite this work, you would replace these identity-specific terms, because it is not OK, really, for white folx to use them (Riley Mukavetz and Tekobbe 2022). *Indigenous* is often associated with peoples south of the US-Mexico border, while *Native Americans* is often used for those in the United States, and *First Peoples* or *First Nations* is largely used in Canada. *Aboriginal* is used in Canada and other countries. And these are all fine terms to use. You will primarily hear me use *Indigenous* to describe us all, and that is because borders are colonial fictions. They are arbitrary and drawn across territories and through traditional homelands. There are differences in our experiences, in our interactions with different colonial entities, if not the same colonial enterprise. But we are all Indigenous to the same Turtle Island. Perhaps someday there will be better words to describe us all, words that do not play into a colonizer's game of who was here first. But I am working with what I have now.

The Chapters

INTRODUCTION: DOING STORYTELLING AS EPISTEMOLOGY
In this chapter, I introduce Indigenous storytelling through the example of the story Choctaw Corn Goddess and its many variations. I lay out my original research questions with which I began this project. I discuss Cottom's notion of thickness of meaning and identity and link this thinking with my theorizing about the cultural roles of storytelling. I describe and discuss settler

colonialism and settler capitalism and how those frameworks influence identity construction. I tell the story of my own journey from thinking of myself as a career technologist to becoming an Indigenous scholar and teacher. Through these stories, I establish the framework for all of the research stories that follow in this book.

CHAPTER 1: INDIGENOUS STORYTELLING AND WAYS OF THINKING AND BEING

In this chapter, I describe and locate key concepts in cultural rhetorics, digital rhetoric methods, and Indigenous frameworks and epistemologies. I do definitional work with discourse analysis, rhetorical analysis, theories of rhetorical listening, and rhetorical image analysis. I define Indigenous notions of good relations, storytelling, ceremony, intuition, and relationality. I explicate storytelling as an Indigenous research methodology and frame out the case studies as research stories. I work with my notions of layered thick identities, rich in context, which Indigenous methodologies are uniquely situated to uncover and value.

Thick Indigenous knowing is a phrase I am using to describe the way Indigenous folx make meaning by conceptualizing stories, art, music, words and terms, ideas, songs, and teachings not as one flat way of knowing but as thick layers of nuanced knowings. In this chapter, I write about thickness and thick knowing in conversation with Tressie McMillan Cottom's notion of thickness. Cottom, when writing about personal essays and personal storytelling, selects and then elaborates on ethnographic notions of thickness, describing thickness as contents that do not fit in the expected and designated spaces (2019, 25). For example, she calls herself thick, because the expectation is that she must conform to white standards of beauty that favor being thin (6). This claiming of thickness in the face of expected thinness here is a resistance to being reduced to a flat interpretation of a stereotypical Black woman. Ethnographer Clifford Geertz explained a "thick" description for sociologists and anthropologists as the ethnographic practice of retaining context when observing culture so that interpretations of culture retain their thickness of meaning and the layers through which meaning is constructed (Geertz 1973). But Cottom carries this idea further in claiming thickness for the speaker rather than the observer, writing, "By interrogating my social location with a careful eye on thick description that moves between empirics and narrative, I have... tried to explore what our selves say about our society. Along the way, I have shared parts of myself, my history, and my identity to make social theory concrete" (26). In the theoretical framing of these case studies, I am combining culture, narrative, empirical data, and identity to describe Indigenous meaning-making.

Here, I use thickness to discuss the ways that Indigenous knowing does not squeeze itself into western knowing. In the chapter, I tell my story of the Chahta Fox as a character in storytelling, a relation, and a biological entity. Fox can be ascribed voice and motives just as easily as she can be classified into taxonomic ranks. Fox is both a creature and a cousin. Fox is all of these notions bound thickly together in the knowing of Fox. I argue that this thick knowing is a distinct difference between Indigenous knowledges and western knowledges, because a goal of western meaning-making is to settle knowledge, while, in Indigenous rhetorics, knowings are not settled but flexible and changeable—they do not simply fit one narrow way of knowing. In this chapter, I write the framework in which the other chapters grow and are scaffolded. I think here of the three sisters, the companion gardening of Natives where our core crops of corn, beans, and squash lift each other up.

CHAPTER 2: LISTEN: SURVIVANCE AND DECOLONIALISM AS METHOD IN THINKING ABOUT DIGITAL ACTIVISM

In this chapter, I take chapter 1 and apply that definitional work and rhetorical scaffolding as methodology. To accomplish this, I explicate a case study from the #MeToo movement and analyze it within the digital Indigenous framework as a working example of doing Indigenous digital rhetoric. The primary approach of storytelling as methodology serves to decentralize western ways of knowing and to subvert western styles of confrontational and oppositional argumentation and evidence. There is also feminist methodology in practice here. Thick meaning comes into play in a discussion of the many outcomes of #MeToo—positive, negative, supportive, purposeful, and otherwise—meaning that it is not necessary to know #MeToo one way, to settle the meaning and outcomes of the #MeToo movement. Rather #MeToo is complex and can be many things collected in a thick, Indigenous understanding of the storytelling of #MeToo and the surrounding responses.

I wrote chapter 2 as a demonstration of Indigenous rhetorics, and as such it is both an exhibit of how Indigenous and decolonial moves operate as well as a case study of how these moves and methods can be applied to an online protest movement. I approach this from storytelling practices, Indigenous identity practices, and collaborative meaning-making. My case study of the #MeToo movement is a practical demonstration of using Indigenous methods to persuade an audience that #MeToo, at its core, is a failure to accept women's stories and storytelling as evidence. In some ways, #MeToo is a movement in which it is only possible for privileged people to participate. I also discuss this chapter's origins as a plenary panel talk that I gave to the 2017 Association of Internet Researchers (AoIR) conference in Tartu, Estonia. The movement of this piece, from observation, to discussions, to analysis, to presentation, through more conversations, to methodological piece, to finally this book, is an example of how knowledge can be co-constructed across time and communities and be an inclusive kind of knowing with our relations. Finally, this chapter is practice rather than presentation, in that I practice what I am describing while I am describing it. It is my hope that the structure of this chapter is useful to students learning Indigenous methodologies.

CHAPTER 3: SKODEN: INDIGENOUS IDENTITY CONSTRUCTION THROUGH FACEBOOK MEMES

I was introduced to the Facebook group Rezzy Red Proletariat Memes by an Indigenous friend who reposted several of their memes to his Facebook wall. I was immediately taken by the raw anger of the posts. In terms of Indigenous identity-making, I have seen a lot of sadness, grief, and trauma. Trauma is discussed among Native Americans as generational trauma, the product of hundreds of years of genocide in the dressing of assimilation, by removals, relocations, reeducation, erasure, and silence. Any anger I knew, I associated with the American Indian rights movements of the 1960s and 1970s; I tended to think, as I had been taught by a colonial education system, that the American Indian Movement (AIM) was a product of the greater era of civil unrest, somehow intimately bound up in student protests, Vietnam war protests, women's rights protests, civil rights protests, and so on. Of course, this makes no sense if you examine it with a lens of survivance: AIM is (because AIM still exists) a separate movement with specific issues to address against the long history of the United States government. It was not a war protest, or an anti-capitalist protest, and it has more in common with the "Long Civil Rights Movement" (Hall 2005) than it does with any 1960s civil unrest. Coming across this group was one of those moments of rupture for me, moments I continue to find as I attempt to decolonize myself. Western schools had taught me AIM was a footnote in an era rather than a movement of its own. RezzyRed Proletariat Memes caused me to confront my miseducation and dig into Indian Country history as having its own storytellers, with its own identity and resistance work.

This chapter, then, is a case study of this Facebook page for Indigenous collectivists. To perform this research, I cataloged and collated screen captures, then sorted them, analyzed the types of posts, and then theorized how those posts construct a contemporary Indigenous political identity that is authentic in its anger as well as its grief, trauma, and silence. I rhetorically analyze the images through Indigenous methods of thickness and layering, discussing how they are uniquely Native American as well as anti-capitalist and anti-fascist. I argue that these memes are liberatory and are resistance work against current movements to suppress Indigenous votes, encroach on Indigenous lands, and violate Indigenous rights.

What strikes me about this case study, and what I want readers to take away from this chapter, is the emotion behind these political memes and posts. I want readers to experience Indigenous anger, and I hope that it breaks open the white supremacist paradigm of the crying Indian whose grief is overwhelming as their people fade into horrifying history. Indigenous people are alive and are making identities for themselves as political actors in digital spaces and on digital platforms—for example, on digital networks employing the hashtags #MMIW, #MMIWG, and #MMIWC (missing and murdered Indigenous women, missing and murdered Indigenous women and girls, and missing and murdered Indigenous women and children, respectively) to organize and share information to not only help find missing persons but also apply pressure to governing bodies to investigate and coordinate to save more lives. The activists using these hashtags demand governments take their concerns seriously and enact legislation to improve the investigations and prosecutions of these cases. The hashtag #IdleNoMore speaks to the resurgence in Indigenous activism, particularly the activism of Indigenous women. And the hashtags #NoDAPL and #StandingRock, among others, were deployed in order to organize and coordinate the protest to stop the installation of the Dakota Access Pipeline (DAPL) across sacred Indigenous lands and waterways. I hope to investigate more of these protest movements in the future, but for now, I focus on one Facebook group to explore one discrete example of digital identity work and activism. This case has been a test of my framework, and hopefully it opens up more research possibilities in digital indigeneity and activist movements.

CHAPTER 4: JEFFREY VEREGGE: A STORY OF RELATIONS

Jeffrey Veregge is a Native American artist whose work came to national and then international recognition when he began illustrating comic books. Comic book art, nerd culture, and pop culture are not the totality of his work; he has produced many pieces for Native communities and programs, for public works, and for his own artistic drive. His art intersects pop-cultural images with Indigenous styles and methods. He is an activist whose work reminds us that Indigenous peoples are engaged in modern life while they also maintain traditional values, beliefs, and arts. Our conversation follows an unsurprising (to me) traditional Indigenous introduction, one where we exchange family, tribal, and linear identities in our opening to our interview, something Veregge and I slipped into by habit, history, and traditional practice, that decolonial and survivance scholars explain and describe in their own works. In an *"Inception"* moment, we were living and being in the means described by the theory and practice of research this book collects and constructs in a holistic theory of digital indigeneity. This interview also discusses how Veregge negotiates a space for himself where he has appeal to wide and varied audiences while still being true to his own people and history. There is discussion about his process and his development, as well as how Veregge uses social media to build community with his audience and disseminate his activist message. We also discuss downsides to social media, some that we all experience and some that are uniquely related to making and representing in Indian Country.

I have wanted to interview Veregge for quite some time, ever since I first saw his Native American interpretations of comic book characters, Batman specifically, which were posted in *Gizmodo*'s popular-culture website io9.com in 2013. Veregge's work takes the present and relocates it on the continuum of Indigenous artistry. This is important, because his blend of pop culture and traditional culture makes visible to a wider audience that Indigenous peoples are alive and making art in the present. I follow Veregge on social media, and through his work, I see his efforts to promote Indigenous causes like education, clean water, respect for the environment, and support for young people. I thought Veregge would be a good choice to fill in the gaps between the case studies I found, with Veregge in his own words explaining how his work and his digital presence function as Indigenous survivance.

CHAPTER 5: MAZACOIN: DECOLONIZING A COLONIAL FANTASY
In this chapter, I describe the beginnings, successes, and ultimate failures of the first Indigenous cryptocurrency, MazaCoin, a Bitcoin variant launched for the benefit of the Oglala Lakota people. MazaCoin and its founder, Payu Harris, captured international digital media attention by attempting to integrate cryptocurrency and affect capitalism on the Pine Ridge Indian

reservation, arguably the most studied Native American population in North America. That digital media attention, with its virality and spreadability, reappropriated the implementation of MazaCoin into western colonial fantasies similar to the nation-making myths of the noble savage and his solitary battle to save his people. This media support, and then rejection, led to the ultimate failure of the project. Here, I examine the storytelling of an Indigenous person and the storytelling of western journalists as a way to demonstrate the distinct differences in approach and intention in Indigenous practices and western practices. The story is unique too in that when I tell it, I include the layers of Payu Harris and the questions surrounding his Indianness both by the problematic white western media and by the more skeptical and nuanced news-telling across Indian Country. In my chapter, Harris is an agent of his own intentions, both colonial and decolonial, when he comments on his own stereotypical Indianness and when he resists such stereotypes by seeking financial independence for his people. When Harris makes references to collectivism and communal strength and resilience, he does so in contrast to the introduction of neoliberal currencies in an already financially fraught space damaged by centuries of settler colonialism and capitalism. In the online media industry, Payu is either this or that: a fraud or a hero, a scam artist or a businessman, or an Oglala Lakota or an imposter. In Payu's own story, however, he can be any of these things together, with all the tension implied, without being forced into any one thing. The thickness of Indigenous stories is expansive and inclusive; Indigenous journalism is not about finding a single answer but existing in a plurality of knowings.

This chapter opens with the story of doing research across a sustained relationship with my collaborator and friend, John Carter McKnight, who at the time was in a postdoc position at University of Leicester studying alt-currencies. My interest in alt-currencies was, I thought, a personal one—I'm interested in culturally situated technologies, and I have been since I was a child playing video games in my family room. It never occurred to me that there might be wide research possibilities in Bitcoin, cryptocurrencies, and peer-to-peer banking back in 2016. I am writing this chapter as a pandemic is emerging, COVID-19, and there are many global digital media stories of small communities using peer-to-peer banking to keep their communities afloat at this time. These stories lack the same romantic overtones that the stories about Payu Harris and his Indigenous MazaCoin have. I think there is room for further study than I offer here, particularly after the full force of the pandemic is measured in terms of its permanent shifts in our society. What will local economies look like after we declare COVID-19 "over"? Since a version of this work was published in 2016, it has been cited multiple times in alt-finance and Bitcoin-specific scholarly papers. When I cowrote the piece that this chapter is adapted from, I did not at the time realize we were writing one of the earlier scholarly works on cryptocurrencies. I am still learning what that means for my own story as I see my work resonating across multiple scholarly fields.

CONCLUSION

This chapter finalizes the findings related to my research questions as well as identifies additional questions I would like to explore in the future. It discusses the relationship between methodology, theory, and practice that is uniquely Indigenous, as applied to digital contexts. It reinstantiates storytelling as methodology and summarizes the book as a collection of research stories. It affirms the vocabulary I have been finding, theorizing, and writing. It discusses potential for further research in Indigenous digital practices.

Taken together, these chapters capture a picture of how Indigenous folx interact with digital technologies. They also document my journey as a scholar shifting from a feminist who researches marginalization and objectification in digital and networked space to an Indigenous scholar addressing a much broader scope of what it means to be othered, what it means to be written out of narratives, and how marginalized folx can and do interrupt these othering and erasing narratives.

1

Indigenous Storytelling and Ways of Thinking and Being

The Hunter and the Alligator
One winter, the village hunters all went out to see whether they might get some deer to bring home. All the hunters but one successfully brought down some fine deer, which they took back to the village. However, one hunter was not so lucky. He tried shooting the deer he saw, but every arrow went amiss. He wandered in the forest for three days and caught nothing to bring home.

Near the end of the third day he decided to give up and go back to his village.

No sooner had he started his journey than he heard an [sic] strange, raspy voice say, "Please help me."

The hunter looked around but could not see anyone.

He turned back to his path home, but then the voice said, "Please, do not leave me here. I'll die. Help me."

"Where are you?" the hunter asked.

"Over here."

The hunter went toward the sound of the voice and soon he came across an alligator. The alligator looked very ill and weak, with very dry skin.

"Please help me," the alligator said. "I need to get to water soon, or I will die. Is there any water nearby?"

"Oh yes," the hunter replied. "There is a nice river off that way, through the forest."

"Can you carry me there? I am too weak and sick to walk all that way myself."

"No, I don't think I should do that."

"I will not eat you. I won't even bite. Not one little nibble. Just please, please carry me to the water, or I will surely die."

The hunter looked at the alligator, and although he felt very sorry for it, he still was not sure whether he could trust the animal's word. But then he hit upon an idea.

"I will carry you on one condition," the hunter said.

"Name it," the alligator said.

"You must let me tie up your feet so that your claws cannot scratch me. And you must let me tie up your mouth, so you can't bite me."

"That is fair. I promised I won't bite or scratch."

The hunter cut some vines, and with some of them, he tied the alligator's jaws tight shut.

Then he tied the alligator's feet. The alligator made no protest at all; it just lay there patiently while the hunter tied it up.

"There," the hunter said. "Now I will carry you to the river. I'll untie you when we get there. I expect you to keep your word that you won't hurt me because if you try, I'll have to kill you, and I'd rather not do that."

The alligator made a noise the hunter took for assent and then hoisted the animal up onto his shoulders and walked to the river. At the riverbank, he gently put the alligator down and cut the bonds on its feet and mouth with his knife. True to its word, the alligator did not try to bite or scratch but rather slithered into the water. It dove beneath the surface and then came up again. It dove and surfaced three more times and then went down again and stayed down for a long time. Just as the hunter was about to turn to go home, the alligator came back up.

"Wait!" the alligator said. "You saved my life. I cannot let you leave without a gift. I see that you have been out hunting. If you do what I say, your family will always have plenty to eat. Go into the forest, and when you see a small doe, do not shoot it. Next, you will see a large doe and then a small buck, but do not shoot either of them. Last you will see a very large buck. Shoot it and bring it home to your village."

Then the alligator slid back into the water, and the hunter never saw it again.

The hunter started on his journey back home, hunting as he went. He saw a small doe, but he did not shoot it. Then he saw a large doe, but he did not shoot it either. Not long afterward, he saw a small buck, but he let that one go too. Finally, he came across a large buck. He nocked an arrow to the string and took a shot. The deer went down, killed instantly. The hunter brough [sic] the buck home, and from that day forward, he never returned to his village empty-handed, and his family always had plenty to eat. (Clayton 2020)

In this chapter, I contextualize this monograph's theoretical work as Indigenous storytelling. I discuss and describe Indigenous storytelling as a rhetorical practice that employs layered meaning, merged with its own cultural context and its own contradictions and variations. I define some key terms. I discuss cultural rhetorics, Indigenous practices, and intersectional feminisms. I discuss the settler colonialism and heteropatriarchy baked into the foundations of our country. Finally, I describe how a matrix of these theories, assembled, works to make visible the complexities of digital identities and rhetorical work of Indigenous peoples.

Indigenous activism and knowledge networks online are not the only application for digital Indigenous methodologies and theories. Unlike western models of knowledge production that center the individual, Indigenous models center knowledge production in collaborative, reciprocal relationships (Arola 2017). This shift in focus can be a helpful tool in understanding how shared digital knowledge emerges and is constructed in collaborative digital networks. It is important to note here that as a digital rhetorician, I am writing about networks in two ways. I am writing about network theory in that I consider connections and symmetrical and asymmetrical relationships between nodes and how these organizational structures help us make sense of digital relationships, including those between humans and other humans, humans and nonhuman actors, humans and software, humans and hardware, and the ever-steepening curve of technological creation and recursion (Galloway and Thacker 2007). Even as I do this, I am also talking about kin and relational networks that are central to Indigenous worldviews expressing that all are kin—we are all relations. That we are all relations is perhaps one of the more commonly known beliefs in the foundations of Indigenous lifeways. I am discussing theory in this book, but I do not write about Indigenous rhetorics and lifeways as theory, because theory is related to the hypothesizing, classifying, and settling of western knowledge matrices. Instead, when I write about Indigenous rhetorics and lifeways, I am writing about knowings. I aim to demonstrate knowing and being, how knowings are made within relations, what Indigenous people know, and the knowings upon which these lifeways are constructed.

I am not theorizing that we are all relations, as in the work of Haraway (2016), as she thinks and writes deeply about kinship networks in the Anthropocene. Although Haraway does write of Indigenous peoples and their contributions to her thinking, and that all "critters" in the Anthropocene are related, unlike Haraway, I *know* we are all relations. And I came to that knowing as part of my lifeway in the process of growing and learning in an Indigenous kin network. Which is why, when I conceptualized this book project, I located it in cultural rhetorics, because it is one thing to say that the Salt River that runs through the town that I grew up in is a relation, an abstraction representing that we are all participants in an ecosystem. It is another to know, as I have come to know, the Salt River as a living companion in my lifeway. The field of cultural rhetorics contends that knowledge production is culturally situated and produced, and it can provide helpful language and frameworks to think about context in online research beyond the nebulous notion of "community." And finally, the power to arbitrate what is and is not valuable is typically and persistently retained by a hegemony of white, male "makers" online. This hegemony tends to reconstitute and reinscribe a rather uniform identity, or a uniform standard for being and knowing, in online discourses and contexts. Intersectional feminisms, with their analysis of a multiplicity of power axes, provide tools for disrupting this monolithic identity, and Indigenous feminisms, with their emphasis on unsettling the settler-colonial state, can fracture the hegemonic heteropatriarchy of digital space. If we are going to see more diversity of ideas and a more egalitarian distribution of identities online, it is imperative that internet researchers find ways to upend the notion of the "typical user," and these tools can assist in that effort.

Storytelling as Framework

I will begin first with the field of cultural rhetorics as my methodological and theoretical home. As Powell and colleagues write, "the project of cultural rhetorics is, generally, to emphasize rhetorics as always-already cultural and cultures as persistently rhetorical. In practice, cultural rhetorics scholars investigate and understand meaning-making as it is situated in specific cultural communities. And when we say, 'cultural communities,' we mean any place/space where groups organize under a set of shared beliefs and practices—American Indian communities, workplace communities, digital communities, crafting communities, etc." (2014). "Communities" can sometimes be a vexing term, given that it can be difficult to define the boundaries and cohesion that creates a community. What makes a community a community? Is a Facebook page a community because it has followers and moderators? Is a group of tweets that share a common activist hashtag a community? What is the difference between an affinity space (Gee 2009) and a community, and how is this distinction realized in literacy practices? This last question is especially significant given that memes, for example, can hold together by affinity group, but affinities alone do not assure that all memers have internal cohesion as an identity group (Noble 2018). Memes are one kind of digital identity-building I explore in this book, and they are commonly described as products of affinity groups shared in affinity spaces (Knobel and Lankshear 2007). As James Paul Gee (2009, 225) writes, "in an affinity space, people relate to each other primarily in terms of common interests, endeavors, goals, or practices, not primarily in terms of race, gender, age, disability, or social class," suggesting that cultural-rhetorical spaces are not affinity spaces, because the affinity space is porous and reorganizing as members enter and leave, or positions change, within the group functionally.

Cultural rhetorics knows that community is, as Powell and coauthors write, groups organized under shared beliefs and practices. In cultural rhetorics, functional purpose is not implied here by the word *community*. For Powell and colleagues, "the practice of story is integral to doing cultural rhetorics. The way we say it—if you're not practicing story, you're doing it wrong. Or, in traditional academic discourse: our primary methodology . . . is to tell stories" (2014). Similarly, Tribal critical race theorist (TribalCrit) Bryan Brayboy writes that our stories are our theories, and that "locating theory as something absent from stories and practices is problematic in many Indigenous communities and in the work of anthropologists who seek to represent Indigenous communities" (2005). For Indigenous peoples, our stories are how we see and think about the world. We tell stories for many reasons: as a way to convey history, as a method for teaching, as entertainment and ceremony, as creative work, as collaborative meaning-making, and for many more reasons. We tell stories to share knowing and to extend imagination beyond what we can observe. Some stories are bound by seasons, some are told only at the right time or the right place, some are only told privately as part of ceremony or commemoration, and some are told publicly, fused with the meanings of who we are, as in the Chahta story of the creation event, where the Chahta came from within the mounds and into the world. Some stories are only told by specific storytellers, who are keepers of their sacred stories. When Thomas King writes that "the truth about stories is that's all we are," he is not just talking about Indigenous peoples but about everyone (2008, 154). King, Brayboy, and Powell and coauthors make a distinction that Indigenous folx are their stories, because stories are central to our identities and are culturally located knowings. These are the same kinds of stories that settlers identified as illogical, not in line with western reasoning or religion, and that served as an excuse to enact cultural violence upon the Indigenous peoples, confirming the western notion of the "savage" for simply not being the same as, or following the same values as, settlers.

At the beginning of the first chapter of this book, I shared a short Choctaw story of the gift of corn. In my mother's files, I found a longer version called "Corn—A Choctaw Legend:"

> In the days of many moons ago, two Choctaw hunters were encamped for the night in the swamps of the bend of the Alabama River. Two hunters, having been unsuccessful in the chase of meat on the preceding day, found themselves on that night with nothing with which to satisfy their cravings of hunger except a black hawk which they had shot with an arrow. Sad reflections filled their hearts as they thought of their sad disappointments and of their suffering families at home. The gloomy future spread over them in its dark pall of despondency, serving to render them unhappy indeed.
>
> They cooked the hawk and sat down to partake of their poor and scanty supper, when their attention was drawn from their gloomy forebodings by the low but distinct tones, strange yet soft and plaintive as the melancholy notes of the dove, produced by what they were unable to even conjecture.
>
> At different intervals it broke the deep silence of the early night with its seemingly muffled notes of woe: and as the nearly full-orbed moon slowly ascended the eastern sky the strange sounds became more frequent and distinct.
>
> With their dilated eyes and fluttering hearts they looked up and down the river to learn whence the sounds proceeded, but no object except the sandy shores glittering in the moonlight greeted their eyes. The dark waters of the river seemed alone to give response in murmuring tones to the strange notes that continued to float upon the night air from a direction they could not definitely locate; but happening to look behind them in the direction opposite the moon they saw a woman of wonderful beauty standing up on a mound a few rods distant. Like an illuminated shadow, she had suddenly appeared out of the moonlit forest. She was loosely clad in snow-white raiment, and bore in the folds of her drapery a wreath of fragrant flowers. She beckoned them to approach, while she seemed surrounded by a halo of light that gave her a supernatural appearance.

Their imagination now influences them to believe her to be the Great Spirit of their nation, and that the flowers she bore were representatives of loved ones who had passed from earth to bloom in the Spirit Land.

The mystery was solved. As one they approached the spot where she stood and offered their assistance in any way they could be of service to her. She replied she was very hungry, whereupon one of them ran and brought the roasted hawk and handed it to her. She accepted it with grateful thanks; but, after eating a small portion of it, she handed the remainder back to them replying that she would remember their kindness when she returned to her home in the happy hunting grounds of her father, who was Shilup Chitoh Osh—The Great Spirit of the Choctaws. She then told them that when the next mid-summer moon should come they must meet her at the mound upon which she was standing. She then bade them an affectionate adieu, and was at once borne away upon a gentle breeze, and, mysteriously as she came, so she disappeared. The two hunters returned to their camp for the night and early next morning sought their homes, but kept the strange incident to themselves, a profound secret.

When the designated time rolled around the mid-summer full moon found the two hunters at the foot of the mound but Ohoyo Chisba Osh was nowhere to be seen.

Then remembering she told them they must come to the very spot where she was then standing, they at once ascended the mound and found it covered with a strange plant, which yielded excellent food. It was ever afterwards cultivated by the Choctaws and named by them Tunchi[1] (corn).

I found this story in my mother's file, and it was originally printed from the Choctaw Nation of Oklahoma website on October 24, 2001. My mother did not have a computer, printer, or internet access, but I have a memory that my cousin Brenda may have printed these pages out so my mother could read them. Or perhaps my aunt Esther Sue. Interestingly, there is a note written on the story that it was "reprinted from October 1992 *Biskinik*," which is the newspaper of the Nation. *Biskinik* online archives only date back to 2001, and the Internet Archive does not have this internet page. One day, I hope to visit the archives in person to read through the old issues and hopefully find more stories. Also interesting is the use of the French *adieu* when the corn lady leaves the brothers. This seems to be a remnant from an earlier translator who was familiar with French or French culture. It is impossible to know where this word came from, but it is evidence of the story changing with the person telling it.

Settler Colonialism

Unlike settler knowledge practices, Indigenous stories are not reified as fact; rather, stories are fluid, layered knowledge practices. Of stories, King writes:

> There is a story I know. It's about the earth and how it floats in space on the back of a turtle. I've heard this story many times, and each time someone tells the story, it changes. Sometimes the change is simply in the voice of the storyteller. Sometimes the change is in the details. Sometimes in the order of events. Other times it's the dialogue or the response of the audience. But in all the tellings of all the tellers, the earth never leaves the turtle's back. And the turtle never swims away. (2008, 121)

Here, the earth on the back of the Turtle, is where we find the knowing of the Americas as "Turtle Island." However, *Turtle Island*, I argue here, is not the fact that Indigenous people's land rests atop a turtle. The knowing of Turtle Island is not a simple creation story. The story also includes all of the variations of the story, variations of storyteller, variations of audience, variations of responses, variations of tone, variations of purpose, variations of peoples, and variations of times. The teller might be teaching a lesson on our origins, when we built mound structures, for example; or the teller might be pointing to our relational and reciprocal cultural practices, as we are all from somewhere on Turtle Island, and therefore we are all related. Even the idea of the planet as a living and autonomous being, the Turtle, is contained here. All these knowings are retained within *Turtle Island* as meaning-making, where it is not simply the story but the culture embedded in and carried along on the story that makes the knowledge. *Turtle Island* reminds us that we are all relations, we share and value the same continent, we know our land is a gift and we protect the land, we know the Turtle is sacred to our origins, and we defend the sacred. Linda Tuhiwai Smith explains that "intrinsic in storytelling is a focus on dialogue and conversations amongst ourselves as indigenous peoples, to ourselves and for ourselves" (2012, 146). She continues that "familiar characters can be invested with the qualities of an individual or can be used to invoke a set of shared understandings and histories" (146). Stories convey the shared understandings and histories that are culture. *Turtle Island* is a piece of our shared culture.

As Wilson describes, methodology is "the science of finding things out" (2008, 34). Above, I described the western scientific classification system as a way to know creatures. Newly "discovered" creatures are identified, described, defined, classified, and settled. Ontologically here, there is one animal kingdom, and the research methodologies are the classification system to know that kingdom. The classification system, taxonomy, is by its nature hierarchal. If we agree upon the western reality, we know how to know creatures. If we view a different reality, the Chahta reality for example, we as researchers will consider our own positionality and subjectivity to examine the reality. Words like "myth" and "legend" are applied by settler-colonial researchers to these stories, and then they are relegated to a historical past where they are preserved as artifacts of a culture, even if the culture is living and breathing and making meaning today. The dehydrated alligator that speaks to the hunter at the beginning of this chapter may be mythical, but he is not a myth. He is a way of knowing about conservation, generosity, reciprocity, relations, and compassion. I argue here that given this cultural knowing, Indigenous reality cannot reasonably be measured as western reality, primarily because knowing and knowledges are not settled, but also because knowings and knowledges are arrived at collaboratively. *Corn Goddess* is a knowing that I incorporate into my understanding of my reality; I can also acknowledge that I know that corn or maize is classified as *Zea mays* in a hierarchy of what I know. I can incorporate that classification into my notion of *Corn Goddess*. Settlers may regard *Corn Goddess* as a mere myth of an ancient people, a way to blur the nagging presence of diverse and diffuse Native peoples with their own worldviews. But *Corn Goddess* cannot be distilled to mere *Zea mays*, because she feeds the people. She is a relation described in our lifeways teaching us the value of corn and reciprocity, and teaching the histories of corn to the people.

In terms of Indigenous meaning-making, the word *worldview* is more apt and accessible than the word *reality*. I prefer *worldview* in part because *worldview* acknowledges the thick and layered knowings of Indigenous knowledge-making practices. It also acknowledges the placing of the authority of their own knowledges in the hands of Indigenous folx themselves, rather than in the quantitative and qualitative methods of the researcher-observer—the sole reality as described by science. It acknowledges the collaborative ties among Indigenous people, ties between their co-constructed knowings and their worldview. This is, as Powell and coauthors describe, culturally situated meaning-making. And in Indigenous storytelling, that meaning is not a simple one of truth-telling but of many aspects such as discourses, dialogues, significance, historical memory, collective memory, and more. And as Margaret Kovach explains, stories are not and have never been simply oral tradition; rather, they are oral stories, as well as symbols, songs, and craftings of many types: "Indigenous people versed in their culture know that sharing a story in research situates it within a collective memory.... A researcher assumes a responsibility that the story shared will be treated with the respect it deserves in acknowledgment of the relationship from which it emerges" (2009, 97). Therefore, to perform research, a researcher must, if not assume the situated worldview, at least acknowledge that the situated worldview is of merit to know, study, and protect.

My research methodology, my storytelling, does not intend to compare Indigenous systems of thought to dominant systems of thought. I am not explaining how one is better than the other, nor do I believe that my research can be used in these comparisons. There is no need to "prove" that Indigenous methodologies, such as storytelling, have merit or can be placed in relation to dominant research methodologies as a measure of worthiness. This book does not represent a rescue or recovery process. Rather, I write of Indigenous methodology because there is limited research about how Indigenous people make knowledge that is unique to us, and the research about how Indigenous people make knowledge and meaning in digital spaces is sparse (Niezen 2005). Knowing that meaning is not fixed is central to our understanding of Indigenous identity-making, and it can give us another theoretical frame through which to examine digital work that is non-western, a contribution to intersectional research.

In *Research Is Ceremony*, Shaw Wilson writes that "research is ceremony.... The purpose of ceremony is to build stronger relationships or bridge the distances between our cosmos and us. The research that we do as Indigenous people is a ceremony that allows us a raised level of consciousness and insight into our world. Through going forward together with open minds and good hearts we have uncovered the nature of this ceremony" (2008, 137). When Indigenous folx say that we want to proceed in "good relations," we are talking about those "open minds and good hearts," a willingness to build a relationship in a mutual exchange of knowing. These relationships we make are imbued with responsibility to reflect and respect the research. Indigenous research is relational, subjective, personal, and emotional and intuitive. It has the potential to respond to the overarching concern that academic research is so grounded in western notions of knowledge that we inadvertently reinscribe hegemonic structures over our research partners and their experiences. Social media is social first, and most of our digital methods do not have a way to account for the social (emotional, intuitive, personal) aspects of digital artifacts. I propose that Indigenous meaning-making, with its fluidity and deep layering, can account for the breadth of engagement and collaboration that makes up the social experience.

Many digital research projects, and necessarily so, focus on the deep control digital platforms and social media providers have over our data, and thus ourselves. Those projects center on biased algorithms, interfaces, subcultures, and gender and racial bias in both technologies and technology culture, and on the destructive powers of trolling (Noble 2018; Phillips 2015; Wachter-Boettcher 2017). In fact, some of my own work focuses on a technological culture that is designed to marginalize women in its process of creating a technological master-narrative for men—that men are masters and makers of technology, while women are passive consumers of technology (Tekobbe 2013, 2015). I strongly believe that more research needs to be done in these areas, as well as in the necessity for a lot more transparency on the parts of many digital content providers, social media platforms, and, frankly, any corporation that monetizes personal data. This project, however, moves in a different direction. Here, I look at networked technologies as tools of potential and possibility to facilitate organizing, identity-building, and survivance on the behalf of Indigenous peoples. As such, this is a discussion on the potential for Indigenous democracies, built in part on social media networks.

As it happens in life, it happens online, and scholars across early internet studies through today have moved on from conceptualizing digital space as different than physical spaces. People bring their values from physical space, transferring them wholesale online; and the biases and ideologies (and methodologies and epistemologies) employed in the physical world, we also use in the virtual world. In 2009's *The Ethics of Internet Research*, Heidi A. McKee and James E. Porter ask important questions about whether the online platforms where digital interactions take place are public space or private space (1–9). They ask us to consider whether the social actors on these platforms consider their own posts to be public or private. These are still important questions, but in the years since 2009 when this book was written, platform designers have leaned heavily on creating tools to make interactions in social digital spaces less private and more identifiable. We are all online in 2024, and social platform identities are inextricably intertwined from private identities. In 2012's *Rhetoric Online: The Politics of New Media*, Barbara Warnick and David S. Heineman describe that social networking sites provide research tools, giving examples such as the ability to discover new friends through Facebook based on how Facebook tracks your current friendships, commonalities, and other shared information on your profile page. They write, "The power of identification as a method for self-promotion across social media is something that hasn't been lost on advertisers or political campaign," which is perhaps as potent a prospect now, at the time this book is being written, in the shadow of the 2020 presidential election, where Twitter is the battleground by which Mr. Trump, the Republican presidential candidate, promotes his political positions and opinions. Other researchers in the field of internet studies also make no distinction between online and offline data collection. In a compelling essay on methodology of data collection, Shani Orgad argues that older communication technology research was never bifurcated between online and offline. Orgad cites research in television, telephone, and radio, for example, and writes, "It has become clear to me that the separation between the online and offline cannot be sustained" (2009, 36–37; see also Baym 2010; Markham 2004). My point here is that it is a settled practice to apply rhetorical theory and methodology from offline projects to online projects, because there is negligible difference about how people are in person from how they are online.

I raise the issue (or specter) of self-promotion to talk about the neoliberal economies of knowing and being online. As Marwick writes, "neoliberal economies are ones in which the production and circulation of knowledge predominates, and where knowledge as a product is emphasized" (2015, 78). Knowledge production on social networking and social media sites is made possible by the tools provided by these platforms, such as sharing, resharing, circulating, recirculating, spreading, and remaking. The tools are designed by a technology industry within a neoliberal model based on the idea that the more knowledge production, the better, as more production means more advertising and other forms of monetization—that the circulation of individual knowing is the ultimate goal on the path to expanding monetization models. These platforms provide the space for "a learned status-seeking practice that both reflects the values of the technology scene and is intimately integrated with social media tools" (128). In my case studies—the women engaged in the #MeToo movement, members of Indigenous Facebook groups who share political memes, the digital life of the artist Jeffrey Veregge, and the rise and fall of an Indigenous cryptocurrency along the arc of the online fortunes of its original founder—are all producing knowledge. They make products like tweets, memes, posts, and cryptocurrencies, but they are also engaged in the thick knowledge-making practices of Indigenous identity work. In some cases, as in those of Payu Harris, the founder of the cryptocurrency MazaCoin, and Jeffrey Veregge, the artist who makes his connections through social media, they are engaged in self-promotion, but not mere self-promotion, because of that thick identity work demonstrated by the tensions in their stories between Indigenous and western tellings of their stories. For example, Payu Harris, the creator of the MazaCoin Bitcoin variant found in my case study, is not simply a digital entrepreneur; rather, his identity is bound up in how he describes and positions himself in the story, how the elders and other members of the Oglala Lakota do or do not know him, and how the western press situates and constitutes him into their stories. Harris relies on self-promotion to bring his currency to his people, but that self-promotion is co-opted and corrupted in the neoliberal systems of individualism, settler colonialism, and the specter of "pretendians."

Neoliberalism, and for that matter, technolibertarianism are powerful forces in the technology industry, because the earliest descriptions we have of the concept of the internet are manifestos like John Barlow's 1996 "A Declaration of the Independence of Cyberspace" (Tekobbe 2013) that helped shape notions of how the internet should or should not be governed. Therefore, like many internet researchers and theorists tell us, the internet is an ideological space, built on the ideologies of libertarianism and neoliberalism, and the technology itself cannot be neutral (Wachter-Boettcher 2017, 11–12). Another important ideology that the internet is built on is white supremacy. As Safiya Noble writes, "white supremacy [is] the dominant lens and structure through which sense-making of race online can occur" (2018, 84). She continues, "often, group identity development and recognition in the United States is guided, in part, by ongoing social experiences and interactions, typically organized around race, gender, education, and other social factors that are also ideological in nature" (84). Technology is ideological, built on the ideologies of its makers, and those makers engage in the neoliberal production of the settler state, a production that encodes its own values atop of the people who use the technology, giving us examples of racism in algorithms and in platforms, through these values. Again, taxonomy is hierarchy, and once taxonomy is applied, racism emerges in the valuing of some over others. One example of this settler-state codification of peoples is taken from Facebook, here as described by Wachter-Boettcher that in 2014, Kiowa tribal member Shane Creepingbear was locked out of his own Facebook account, because his name was in violation of the Facebook policy that users must use their real names. Facebook flagged Creepingbear as not a real name, with the Facebook message

You Name Wasn't Approved.
  It looks like the name violates our name standards. You can enter an updated name again in 1 minute. To make sure the updated name complies with our policies, please read more about what names are allowed on Facebook. (2017, 53)

Facebook offered no other options and no other way to confirm that Creepingbear was his real name. Wachter-Boettcher also writes that several other Native American names were rejected by Facebook, including Robin Kills the Enemy, Dana Lone Hill, and Lance Brown Eyes. Of the incident, Wachter-Boettcher cites Creepingbear as saying that "the removal of American Indians from Facebook is part of a larger history of removing, excluding, and exiling American Indians from public life, and public space" (54). Creepingbear says this policy aligns with centuries of Native culture erasure. In short, Facebook's coding requires users to use their real names, and yet does not accept the names of some Indigenous Americans as "real." Why not? Because those who wrote the code for determining what is a "real" name did not imagine the names of Natives. Without serious reconsideration of what constitutes a name, Native Americans remain unnamed by the settler logics of Facebook and other online platforms that require a specifically western naming paradigm, disenfranchised from the global digital community.

Telling our stories challenges white supremacy because it makes visible who we are. As I discussed earlier, thick meanings and thick knowings ride along on our stories. Our very identity work is made of these complex, thick layers that do not align one-to-one with western knowledge practices or conventional data-coding practices. I argue here that without thinking about the digital products of Indigenous peoples within the thick meaning framework, the assumptions of white supremacy are replicated and reconstituted in digital, Indigenous research. I share here a story about the context of chapter 5, in which Payu Harris and the cryptocurrency MazaCoin that he launched are discussed. John Carter McKnight and I originally published a version of that chapter in 2016 on the site of the internet research journal *First Monday*. While I was going back and forth with the editor to finalize project, the editor and I came to politely disagree about a fine point in the piece—whether or not Harris's description of himself as a member of a historic legacy of the Oglala Lakota, who have to this day never surrendered to the United States government, caused in some way the kinds of journalistic embellishments that eventually led to the failure of Harris's project. That coauthored article originally focused on the cryptocurrency and the potential for alternative currencies among marginalized and nondominant groups. In the end, McKnight and I ascribed the failure of the Indigenous cryptocurrency at the time to the perilous nature of instituting an ideologically value-laden currency on the economically precarious Pine Ridge reservation. But I also included rhetorical analysis of the technology and finance sectors' media treatment of Harris as a kind of historical caricature, with stories similar to European traveler's tales. In the article, I wrote that Harris invokes the knowing of the buffalo as a plurality of resources and reminds us of the Oglala Lakota as peoples who have never surrendered to the United States government. The journalist described MazaCoin as Harris's Little Bighorn. In the piece, I called that journalist biased against Indigenous folx and skewed in his representation of Harris. My editor asked me, with good reason and good intentions, if I did not think that Harris brought some of the mistreatment on himself when he invoked the heritage stories of the Oglala Lakota. I disagreed then and I disagree now, because of the complexity and multiplicity of meaning, the thickness, of Harris's imagery and storytelling that cannot by its nature have the western one-to-one relationship with identity. I do think Harris, like so many of us, has absorbed some measure of settler-colonial thought. When we discuss decolonizing ourselves, we are referring to these settler-colonial paradigms—but if Harris sees himself as an Oglala Lakota warrior, then it is a simple matter for journalists to subsume that image into a greater story of a standoff between one man and the federal government. When Harris invokes symbols and stories of his people, he and we, the audience, do not have to accept the distilled single meaning of non-Native journalists. The editor and I worked it out, but the misunderstanding stayed with me and informs how I look at all "romanticized" stories told by settlers about Native American histories, as well as the stories Native Americans tell about themselves. This is where I find thicker meanings.

In chapter 3, I discuss Indigenous identity-making and Facebook memes. I share and describe a number of memes from an Indigenous, socialist-identifying page. While circulation studies discusses how digital artifacts are transmitted across digital spaces, spreading out into common knowledge—for example, how an image macro becomes a meme that keeps showing up in your Facebook newsfeed—the Facebook page I study here is about collecting memes (Gries and Brooke 2018). While I acknowledge the contributions of circulation studies to the understanding of how these memes move from one person to another, to groups, and across platforms, my work with memes is different. The memes are collected and shared to the page of the community that comments on and responds to the texts. This collection process is about constructing a community that is relational in its identity work and reciprocal in its comments and conversations. It is important to note here that this Facebook group has helped shape my own political identity, in that I have found a strong sense of commonality among others who appreciate the bitter humor of those memes. When one of the memes demonstrated that while Mr. Obama's Indian policy was to send cops to the NoDAPL protest, Mr. Trump's policy was to send even more cops, I laughed aloud. I thought to myself that this is always the case when Indians organize—the federal government has historically sent troops to suppress us. This history is unique to Native Americans in that there is historic legislation that offers bounties for our bodies, strips us of our rights, and removes us from our lands, all reinforced by militarized forces. Aspects of this history are shared with other groups that white supremacy wants to suppress. For example, peaceful protest actions in the Civil Rights movement were met with the violence of militarized police forces. Here, institutional violence against people who are raced by white supremacy is a thick knowing for Indigenous folx, Black folx, the Japanese American community, and many other groups as well. It is also woven into the fabric of the narrative of the United States. There is no settling this knowledge as long as white supremacy is the organizing principle of American culture and society. Not unless we unsettle it.

In theorizing identity, I want to again touch briefly on the genocidal practice of assigning percentages of "Indian blood" to Indigenous peoples, described in the introduction. Who is Indian and who is not is determined by a deeply troubling, personal, and painful identity-making history that is bound up in US governmental efforts of genocide. Here, I want to make clear that because of the way blood quantum divides and erases some while privileging others, as is the way of racism, I cannot emotionally distance myself from this discussion and that it is a topic that should always be approached with great sensitivity and ethical consideration. The process of deciding identity based on "blood" alone is death, and so when I write about it, know that I write with the weight of the murder of nations on my shoulders. For my discussion here, it is important to note that when Native Americans were removed from our own lands and communities through the Indian Removal Act, it was closely followed by government efforts to limit the Indigenous population by physical inventory. The Dawes Commission of 1893 began assigning blood quantum during the process of land allotment. There were two major changes here that destabilized tribal society. The first was that the system of land allotment approached the distribution from settler-capitalist and patriarchal frameworks where land was assigned to "full blooded" Indian heads of household, instead of retaining the land in tribal collectivism as had been the prior history of the land and people. Natives were suddenly thrust into capitalism and patriarchy, which fractured their own identities and collective knowings.

The second major change was the idea that identity and tribal belonging could be quantified by blood, or "blood quantum." While there was no way for members of the Dawes Commission to use science or western medicine to identify quantum of blood, instead they relied on western-style observation. In other words, they relied on their own opinions about who was Indian and who was not, largely based on phenotype and willingness to cooperate with the Commission. Where the construct of race had been imposed on Indigenous peoples' identities, suddenly the possibility of losing one's race became a threat to that same identity. And race was intrinsically tied to material assets and conditions, linking the settlers' notions of race with very real-world consequences. If you were Indigenous and a head of household, and if you looked Indian enough and behaved yourself according to the arbitrary standards of the white people, you could be given land and enrollment. If you were not a (male) head of household according to newly introduced white patriarchy, you might be left out. Under these circumstances, identity became a knowing not of who you come from and who claims you but of your distance from ancestors who were "full blooded" and also of how you looked. This material identity has been internalized by Indigenous peoples, and today blood quantum is largely used across Indian Country to identify who can receive tribal rights and benefits and who cannot.

Blood quantum is ever-diminishing. For example, while I have a number of Choctaw relatives in my family tree, and my father's family has family memories of visiting the Cherokee reservation to visit their relations, my tribal registration is tied through my maternal great-grandmother's Dawes registration as "Indian." My mother is, according to blood quantum, "less" Indian than her mother, or "part Indian." I am also, according to blood quantum, "part Indian." If I had children, they would also be "part" Indian, and my grandchildren may not be Indian at all. In blood quantum, the Choctaw in my family disappears in just a handful of generations. This is the colonialism of blood quantum, designed as it is to make us eventually disappear. Kimberly Wieser-Weryackwe offers a nuanced discussion of how being "part" Indian is a colonial construct, in her essay "Aunt Ruby's Little Sister Dances" in the *Unpapered* collection. She notes that her Cherokee and Choctaw ancestry predates the Dawes rolls and thus she is not documented in the way that I am (192–93). She is not, however, less Indian or more Indian. We are both of Indian people who claim us.

As Mikaëla Adams writes in *Who Belongs? Race, Resources, and Tribal Citizenship in the Native South*, Indigenous identity in the American South is particularly complex given that many of the Indigenous folx in the South had long been integrated into community with white settlers and formerly enslaved folx and their descendants. In the South, the Dawes Commission fractured established communities and created confusion for the previously "assimilated." Adams comments here on my own people, the Chahta, who were among the "five civilized tribes" originally exempt from the Indian Removal and Dawes Acts, who were rather quickly included in removal once the government set its eyes on their land: "Denied the privileges of whiteness because of their phenotypically Indian appearance, they turned to their Choctaw relatives for social support. The 'full-blood' Choctaws accepted them based on their kin ties and their familiarity with the Choctaw language and culture. Indeed, as far as other Choctaws were concerned, their degree of 'blood' had very little to do with their identity as Choctaw" (Adams). There are more issues here, in that blood quantum and tribal rights are only considered for "federally recognized" tribes. There are tribes and bands across the country who are not federally recognized and, according to the government, not Indian at all even as they have lived on their lands for centuries or were removed together wholesale to other lands. To put a complex and nuanced issue simply, in this country, you are Indigenous if your ancestors are from a federally recognized tribe and you are a close relation to an already enrolled ancestor. It is a common practice across Indian Country to terminate enrollment for anyone who is more than three generations from a "full blooded," direct ancestor. Prior to the Indian Removal Act and the Dawes Commission, the pieces of people were not considered related to Indigenous identity, which was drawn from, as I wrote above, who your family is and what community claims you as their own. The white supremacist reliance on phenotype, a system of western knowledge-making that catalogs a group by its common physical traits, complicates Indigenous identity further. "Looking" Indian is as persuasive as blood quantum in a white supremacist paradigm. Not looking Indian enough in the subjective opinion of the white observer is taken as authoritative knowledge, as opposed to how a person identifies as belonging.

I also want to note here that as I am discussing the stakes of government-sanctioned Indigenous identity, it is beyond the scope of this project to discuss "pretendians"—pretend Indians who represent themselves with an Indigenous identity for social or material gains. Here I offer the example that my sister and I are enrolled members of the Choctaw Nation, while her adopted children are not. And this is due in large part to white supremacy, which has already assigned various races to her children, but also to hesitation by a tribe to extend rights to those without blood quantum evidence. These children are being raised by their Indigenous mother, with a great deal of engagement with their Native grandmother, aunts, and cousins. They have Indigenous lifeways and worldviews taught to them by us. Outside of the white supremacist, settler-colonial imagination, we are a family living in reciprocal community with each other. But today, that is not enough to be Indian, neither to overcome their own raced assignments, resist their stereotypical phenotypes, nor even claim the culture of their own mother. My sister raises her children with emphasis on their multiculturalism, honoring the identities of their birth families, and they know they are not members of her tribe. Because this is how white supremacy works. The whole issue needs decolonizing.

Importantly, though, in the context of this book, I include this information here to add critical nuance to my discussions of Indigenous identity-making and culture-making. My nieces and nephews do not "look" Indian, because they are not phenotyped in this way. They are of Latinx, Black, and white ancestry. A white supremacist gaze (whether the gazer is white or not) is cast on the raced person: for example, my blue eyes and light skin do not measure up, but my cheekbones and nose hint at a phenotype of a raced identity. My nieces and nephews have brown eyes or blue eyes, or curly hair or straight hair, or light skin or dark skin, and they are always raced by white observers. This is a mess. I ask that you keep all this mess in mind when considering identity, both as I theorize it here and as the folx in these case studies construct it themselves.

There is a great deal of scholarly discussion on what digital identities are and how they are constructed, and then how those identities mirror or reflect those in physical space. I add to that conversation here as I theorize storytelling, meaning-making, lifeways and worldviews, economic ideologies, white supremacy, and settler colonialism in the complexities of Indigenous identity-making. In fact, a lot of discussions about online identity practices are written from an undeclared and default white perspective. I hope I have explained here why this "generic" positionality (that is actually white, settler, patriarchal, and neoliberal) erases race generally and indigeneity specifically.

I think there is much to do in decolonizing and deconstructing the underlying hegemonies that shape online identity without consideration of the nuances these power profiles impose. In the case studies I present here, I hope you will keep in mind that this book is simply a beginning. In the next chapter, I will demonstrate this theorizing in practice as I present my methodologies contextualized in a case study from an earlier AoIR conference I attended.

2
Listen

Survivance and Decolonialism as Method in Thinking about Digital Activism

Why the Turtle Has Cracks on His Back
Choctaws knew Turtle long ago, when the back of his shell was smooth. One evening, Turtle heard loud chattering among the winged forest animals. "What is going on?" he asked.

"We are getting ready to fly south for the winter," a bird answered.

"I want to go with you," said Turtle.

"But, you can't fly," Bird crowed.

After much thought, Turtle came up with a plan to fly South with the birds. "On the day you leave, I will find a sturdy stick," he said. "I will bite in the middle of the stick. Two birds will carry the ends of the stick in their beaks. That way, I can go with you."

On the agreed upon day, Turtle got up very early. He found just the right size stick and went to where the birds gathered. Turtle clamped tightly onto the stick. Two birds held the ends between their beaks and began to rise into the air.

As they flew higher in the sky, crows approached from a distance. "That turtle is flying with the birds," said a crow. "Who thought of such a smart idea?"

Turtle wanted the crows to know that he was the smart one. He opened his mouth and said, "I DID!"

https://doi.org/10.7330/9781646426478.c002

Turtle fell down, down, down. He landed hard and his shell cracked into many pieces. Turtle's shell still shows what happened to him the day he fell from the sky. ("Chim Afvmmi Na Yukpa!" n.d.)

Background

As both a discussion of Indigenous methodologies and a working case study of how these methodologies can be implemented, this chapter is an elaboration and explication of the talk I gave for the plenary panel at the annual conference of the Association of Internet Researchers (AoIR) in Tartu, Estonia, on October 21, 2017.[1] I am weaving methodology into the text of the talk to demonstrate through storytelling how Indigenous methodologies can be applied in practice. I originally gave the talk to introduce to the AoIR audience Indigenous knowledges and cultural practices in exploring both gendered online abuse and online activism against that abuse. My goal was to demonstrate how decolonial and survivance praxis could lend new perspectives and approaches to investigating the important problem of online abuse, and general claims of ineffectiveness in online activism. I argued that the typical western paradigm of research to settle knowledge, or true / not true efforts to prove claims, were inadequate to understand the thick meanings inherent in meaning-making and identity work. In this chapter, I retain those goals and extend them further to describe these methodological approaches and how they work, as I instantiate them. First, in this chapter, I will introduce Indigenous self-identification, storytelling, and thick meaning as a heuristic and as praxis. I will describe and discuss intersectional feminisms, cultural rhetorics, and Indigenous practices, providing some examples. I will describe and discuss intersectional feminisms in relation to white feminism, a kind of feminism that can reinforce the neoliberal enterprise. I will discuss how this neoliberal enterprise can subvert online activism by reinscribing power when it subsumes the activism into the marketplace—this is selling activism rather than using it to rupture dominant discourses, narratives, practices, and beliefs. I will discuss allyship and how survivance and decolonial practices can introduce novel ways of approaching activism. Finally, I will propose a framework for online activism that is conscious of the problems of neoliberalism, white feminism, and the overarching homogenization of identity online.

This chapter is written in first person and is an account of an Indigenous-style collaboration with an audience to share and make knowledges together. This first-person collaborative storytelling is an element of Indigenous style, and there are many more contained in this chapter. While I was the only one speaking from the front during my talk, the collaborative nature here provides for the thick meaning of the emotional responses of the audience, their verbal reactions, their gestures, their tweets, and their interactions with each other. The room was not silent, the audience-participants were not frozen in place, and in addition to interaction with each other, many were texting, tweeting, and taking notes on their electronic devices. AoIR is a technologically embedded conference, so the simultaneous production of digital and personal responses was to be expected. One useful academic reference here on working in Indigenous contexts is Gregory Younging's 2018 *Elements of Indigenous Style*. This text identifies specific style elements for Indigenous writing and provides explanations of ethical practices when working as an Indigenous researcher or collaborating with one. Other useful texts are those on decolonizing research methods, for example, *Decolonizing Methodologies* by Linda Tuhiwai Smith (2012).

It is important to note the *kairotic* moment in which I gave this talk back in 2017. *Kairos* is the rhetorical concept of the significance of time in a rhetorical context. This was the moment when offering a talk on #MeToo would have significant relevance to the audience. On October 15, 2017, American actor Alyssa Milano posted to Twitter a suggestion: "If you've been sexually harassed or assaulted write 'me too' as a reply to this tweet" (Milano 2017). This act kicked off a wave of women, and some men, sharing their sexual harassment and assault experiences on social media, including many actors, musicians, and political figures lending their voices to the movement. When I originally conceived of and wrote the talk, I had intended to use Gamergate and Jezebel.com as case studies to model my Indigenous practices toward analyzing and constructing meaning from the tweets and the cultural responses to those tweets. However, by the time I was on the plane to Estonia, the #MeToo movement had reached critical mass, spreading around the world, a movement that would eventually total more than 19 million tweets (Pew Research Center 2018). In this critical moment, I revised my talk to include this movement that was happening online and in mass media while we were attending AoIR, explicitly because of its immediate relevance to internet researchers, whose annual conference was playing out both in person and on social media platforms like Twitter.

As I write this book chapter in the summer of 2020, I reflect on the cultural impact of #MeToo since 2017. I also contemplate why #MeToo was able to gain the attention and traction that earlier sexual harassment and sexual assault activists had not. First, the 2016 election cycle saw the global Women's March, as well as a record number of female candidates running for public office. I contend that this is, in part, a reaction to the Trump presidency, where an acknowledged sexual harasser was at the top of the Republican ticket and then elected president of the United States.[2] There are many factors to interrogate about the 2016 election, both in the Trump candidacy and the very public defense by powerful men of Trump's misogynistic "locker room" talk and history of abusive relationships with the women in his life. Coming ten months after the inauguration of Donald Trump, #MeToo rose out of the already simmering anger of American women. Today, that anger is still fueling women's activism, as evidenced by the number of diverse women who took office after the 2018 midterms, including Indigenous and Muslim women. So, in this sense, #MeToo is both a movement and a catalyst for continuing women's activism in the run-up to the 2020 election, with a Democratic candidacy pool featuring a historic number of women. #MeToo, even with its problems, which I will discuss later in this piece, is not a movement that began and ended in 2017 but one that launched in 2017 and continues to resonate in 2020.

Survivance: A Story of Many Meanings

Generally, western knowledge is formed in the mind as a product of intellect, while Indigenous knowledge practices are experiential, a co-constructed engagement between people (Wilson 2008). As I wrote in chapter 1 of this text, Indigenous knowing is thick knowing, with stories, gestures, and audience responses woven together. In this chapter, I include in italics the original text of my talk, interwoven with the expected knowledge practices of academic writing, including research questions, theoretical groundings, methodology, and definitions. I also try to answer the poignant question of one colleague, who asked, "What does it mean to insert the tongue and practices of [my] ancestors into an academic research publication?"

I originally designed the talk as an oral piece, as Indigenous storytelling—to be spoken aloud with aural cues, to be heard by an audience, and to be emotionally evocative. I acknowledge the Greco-Roman origins of my field of rhetoric as well, and this talk is a blending of these traditions. My design approaches the talk as *techne*, a performance employing rhetorical affordances of speechcraft, of evoking affect, of inviting audiences to a dialogue, of co-creating meaning with an audience. I also position myself as an Indigenous storyteller, one who experiences meaning-making as a thick practice of multiplicities. The original text, as performed, contained repetition, gaps of silence, breaks in thoughts before completing them, and inflected words to mark dual or multiple meanings. These aural cues were included to make space for audience contemplation and reaction—to create accommodations for audiences of different cultures and practices to join the conversation through the Indigenous practice of "good relations" (Watanabe 2014). Through these cues and in this performance, my intent was to structure my talk as a collaboration between myself, some of the audience as members of marginalized cultural and social groups, and the larger portion of the audience as members of dominant groups and writers of dominant narratives. Here, in this chapter, I attempt to retain some of the performative components of the piece while adapting it and its embedded practices for a reading audience. Here, I emphasize my own Indigenous knowledge practices while attending to western academic writing.

Survivance, as described by Vizenor, is a combining of the survival of Indigenous peoples with resistance to forces of power that would erase those same peoples (1999). It is activism, enacting Indigenous identity in institutional spaces and against hegemonic orientations of power. It is community in that it is the practices Indigenous communities use to preserve and sustain their existence. It includes existing as an identifiably Indigenous person in the presence of others not from the surviving community. Survivance is working toward self-determination through Indigenous knowledge practices. The first knowledge practice I enacted in this talk was that of self-location:

> Indigenous researchers will situate themselves as being of an Indigenous group. . . . They will share their experience with culture, and/or they will identify the Indigenous epistemology (or epistemologies) of their research. Often, they will culturally locate in all three ways. . . . For many Indigenous people, this act is intuitive, launched immediately through the protocol of introductions. It shows respect to the ancestors and allows community to locate us. Situating self implies clarifying one's perspective on the world (Meyer, 2004; Hampton, 1995). This is about being congruent with a knowledge system that tells us that we can only interpret the world from the place of our experience. (Kovach 2009, 110)

I began my talk with a greeting in the Chahta language, the tongue of my ancestors, as a means of marking myself as Indigenous and making my community visible. It also shares something of my culture with the audience. I introduced myself, giving my name and that of my community. I addressed the audience as "beloved relations," *relations* being an Indigenous knowledge that all lives are connected—we are all related. I offered that my audience was beloved in an accommodation of peace, in that I was giving my talk to an audience I respected and whose collaboration I valued.

> Halito!
> Akim achukma?
> Su hohchifo yut Cindy Tekobbe Chahta sio Tvshka Homma.
> Hello, beloved relations. How are you today? I'm Cindy Tekobbe, and I come to you today as a member of the sovereign Chahta Nation of Oklahoma of the Americas.
> I am an intersectional feminist theorist, an Indigenous internet scholar, and a cultural rhetorician.

Here, I identified myself to the audience as a researcher, an Indigenous one, and presented my heuristics—my self-located view of the world—as an intersectional feminist theorist, Indigenous digital humanist, and cultural rhetorician. These are my areas of research and the lenses through which I view the world. Indigeneity is also my worldview. Here, I identified myself culturally in all three ways Kovach articulates above, with my identity group, my culture, and my epistemologies. I did so intuitively, as part of the procedure of introductions that typically precedes an academic talk. My audience, while perhaps not able to identify the specific practices of survivance here, were familiar with the practice of self-introduction because it is commonplace. Through my talk itself, I planned to engage in the defining and demonstrating of these survivance practices to the audience in Tartu:

> *Introducing myself in the tongue of my ancestors is an example of intersectional feminist theory, indigeneity made in practice, and cultural rhetorics enacted.*
> *For the Indigenous and First Peoples of the Americas, using our language that was taken from us in government-sponsored campaigns of genocide, that persist even today, is a practice of survivance. It says, we are a living people. We survive.*
> *I survive.*
> *And it's not just survival. It's embodied practice. Through my language and my Indigenous body in front of you employing this language, I remind you that not only are we a living people, we are among you, resisting the slow death of erasure and silence.*

Here, I first draw attention to survivance as an embodied practice in that I am with a group of people who are listening to me and responding to me, standing in front of them physically, using my ancestors' language verbally, and offering a phrase affirming these realities, "I [we] survive." This is thick meaning—survival as a collaborative performance with my audience as witness to survivance through me, my language, my performance, and my body. Another aspect of speaking in the language of my ancestors is both a survivance practice and a decolonial practice. The colonization of knowledges is one where the colonial researcher assumes the right to define what they observe. People are objects of study, and those people and their cultures are examined through western heuristics of reality, time, and space (L. Smith 2012). In Indigenous knowledge practices, this Chahta language introduction positions me to my audience and describes the practice in which I am engaging—I am speaking, albeit briefly, in the language of my ancestors, the Chahta peoples. Giving one of the talks on the plenary panel at AoIR 2017 was arguably the most attention my scholarship had received to date. I made the choice on this important occasion to deviate from western notions of academic performance, to use the language of my ancestors, because it is deeply significant to me. It is deeply significant to Indigenous peoples, because to speak our languages in institutionalized and public places has been forbidden by North American governments and cultures (Dunbar-Ortiz 2014) in acts and campaigns of genocide. Our traditional voices were silenced, and with them, our traditional ways of making meaning and what it is to know. Here I am engaged in an act of survivance, but also one that is decolonial, because I am an Indigenous woman inserting her heritage language back into the public sphere from where it had previously been erased.

I am still here. I persist.

I said I was a cultural rhetorician, and I'll take a moment to unpack that. A cultural rhetorician studies the ways that culture is employed to imbue things and rituals with meaning—the way cultural practices make knowledge—and that knowledge is situated within and meaningful to specific groups of people. It is their own knowledge and meaning.

I have just demonstrated a cultural rhetorical practice to you with my personal introduction.

My culture knows women to be leaders of families, keepers of history, ethical practitioners of justice, and conveyers of knowledge.

Through the use of my heritage tongue here, I instantiate myself as a Chahta woman, a voice of my family, here to convey to you today that which I know to be true.

That not only am I still here, but through my language, I call this space my own with the blessings of my family and the ancestors who walk with me.

I survive.

I am still here. (I persist.)

> *I am legitimized in this space and place that is my own because I speak it so. This is my agency.*

I pause here for definitional work. I have introduced cultural rhetorics, intersectional feminism, Indigenous rhetorics, and internet research into this talk. As I reiterated, cultural rhetorics is the study of how cultures make knowledge: "All cultural practices are built, shaped, and dismantled based on the encounters people have with one another within and across particular systems of shared belief. In other words, people make things (texts, baskets, performances), people make relationships, people make culture" (Powell et al. 2014). Thus, relationships—human interactions—make reality, as through them our cultures produce the construct within which we exist and process the world around us. And, like Indigenous scholarship centers relations, cultural rhetorics centers relationships. In contrast to many western ways of knowing, in the practice of cultural rhetorics, not only is knowledge not located in the mind of the individual, it is specifically created through interactions with other people. I offer a few examples here, including that my culture knows women to be at the center of families, and that because I am speaking—storytelling—I am constructing a cultural space. To note, I am speaking for myself, as a member of a family, and as a member of a people. Here, in this role, I do not explicitly state that all that I represent is everything my family represents and everything my people represent. I am not the representative of Indigenous people in general but rather within my relations network. This is the balance, while I am collaboratively associated and sustained by all my relations, I do not speak for them. I speak because they sustain me.

While its recorded history describes the Chahta peoples as matriarchal, these women-centered family structures are not abandoned to a historic record but are enacted in everyday Chahta life. My own extended family is currently organized around my mother, with myself, my sister, and my cousins reaching out to her for advice and deferring to her experience on decisions. For example, I sought her counsel on whether or not to use my family's Chahta name, Tekobbe, rather than the patriarchal practice of taking the names of my father and father-in-law. My decision to attend graduate school was made at her kitchen table, and the speech at my graduation party was given by her. When we decided together whether I would continue my schooling, we discussed her concerns that more western education would distance me from my roots. Her phrasing deliberately confronted settler logics, asking, would I, with my graduate work, come to think I was better than my family, that my newly learned ways of being, identifying, and knowing were somehow now better than what I was raised to be. In this moment, she was recalling and invoking the fraught history of Indigenous boarding schools that were meant to "kill the Indian and save the man," but she was also talking about family knowing, the way communication and identity are woven together in domestic meaning-making. Here, we both represent what it means to be Chahta, what it means to be ourselves, what it means to be a people, and how these sustained relationships in everyday spaces help us navigate a reality that is not in step with the way we know and build knowing.

The settler-state and church education of Indigenous peoples has always been an enterprise of erasing the Indigenous and replacing it with the western (Brayboy 2005). In the years since I graduated, I have accepted the responsibility to maintain my family's history and culture, and I have demonstrated my commitment to retaining these knowledges as I continue to learn and practice scholarship. Chahta culture produced what I know about being a daughter, it shaped my career path as an educator, and led me to accept a cultural role of keeping history, a history I share with you here. I hope it is apparent in my example that the way of being a daughter in collaboration with a mother guiding her family is different from the patriarchal structure of the unilateral authority of a father, where the bodies of daughters are a resource for their material and cultural value. I learned *daughter* from family stories, from the modeling of women in my life, and from my mother's treatment of me as an equal, valuable for myself and not what cultural currency I represent. This thick knowing of *daughter* allows me to retain my sense of identity with my mother while performing in the world as scholar. *Daughter* is many things in Chahta life, but unlike western daughters in a settler-colonial context, I am not and never would be family chattel, for example.

Feminist methodologies, like Indigenous methodologies, also expect a researcher to locate herself and make visible her positionality when writing or speaking about her scholarship. Feminism can generally be thought of as the advocacy for equal rights for all. Indigenous feminism has outlined the impact of colonialism on Native American culture and identity (Deer 2015). For intersectional feminist scholars, the term "feminism" alone neglects the complexity of social power and privilege enacted on the bodies and identities of non–dominant narrative writers. Social inequity is shaped not only by a single factor, like gender, race, or class, but alongside many factors interacting and influencing each other (Collins and Bilge 2016). Therefore, intersectionality is both a heuristic and a praxis, theory and practice, linking together and mutually informing one another. I return to my example of *daughter*, growing up in the middle-class, suburban context of the United States. I am what is known as an "urban Indian," one who has not grown up on a reservation. Instead, I grew up in a midsized city in the American Southwest. I did experience *daughter* as the child of an Indigenous mother, but also as the child of an Evangelical Christian father. I experienced *daughter* as defined by conventional American cultural standards, as a girl who went to public schools and Christian churches. And as a girl who knew *daughter* to also mean I was to be protected from the untoward attentions of boys while committing to obedience to my father as the leader of the American nuclear family, even if our family was not the model of nuclear but maintained close relationships with extended family.³

Historically, through the enormous and terrorizing efforts of settlers and colonizers, Evangelical Christianity is the politically dominant religion in the United States. Its conservative arm is particularly influential in United States culture and government. Most legislative action regarding women's bodies and women's autonomy is sourced in this deeply traditional, patriarchal power matrix. While Indigenous North Americans are generally collectivists, the national economy is capitalist, and its political paradigm is neoliberal. Neoliberalism privatizes the state and monetizes its citizens: for example, the widespread privatization of prisons, known as the prison industrial complex, that profits off the mass incarceration of its citizens in relatively precarious positions, such as people of color and the poor. My knowing of *daughter* is not simply the cultural product of my Indigenous mother and western, culturally traditional father. As I describe here, my knowing is also intersected by the axes of collectivism, capitalism, neoliberalism, race, class, gender, settler colonialism, and others—*daughter* is thick with meaning. In other words, the self and the political should be understood as shaped not by one force but by many, intersecting and influencing each other. Intersectionality is a useful analytic for accessing the complexities of a condition or set of conditions and for this type of social justice and intellectual work.

And if something about the foundation of this talk seems somewhat familiar to you, then I am glad that we can make meaning together.

   Because I want to talk to you today about two online cultural rhetorical practices that share some commonalities with my own:

   First, I point to the MeToo hashtag that spilled across social media over the last week. In summary, victims of sexual harassment used #metoo to share their firsthand accounts they have endured and survived in an effort to demonstrate the near universality of the problem. And the results were sobering as tales poured out across the internet from women and nonconforming folx who cut across racial, cultural, economic, and social groups.

However, this chapter is written in the context that patriarchal colonialism is a colonial belief system, and the right of the patriarchal power matrix to control women's bodies and to arbitrate what is true or valuable in relation to women's issues is nearly universal (Spencer-Wood 2016). Speaking and writing in this context, I identify that one problem with reporting and acknowledging sexual harassment and sexual assault is that victim accounts are often not believed or are received with a level of skepticism that can be insurmountable. In chapter 1, I grounded this skepticism in the western analytical oppositional paradigm where something must be false until enough evidence has been offered to demonstrate truth. As our court systems say, innocent until proven guilty. I am not arguing that sexual assault stories should always be seen as true. Instead, I offer that this western knowing is not truly objective, as is typically assumed by the western knowledge matrix, but rather subjective, based upon the normative standards by which the stories and evidence are weighed. In the case of the general structure of patriarchy, women are chaste, men are unable to control their impulses around women, and therefore women must guard their bodies against men. Failure to do so is the failing of female responsibilities, not male character, and certainly not external influences like economic disparity, power imbalances, social precarity, patriarchy coded into marital and family law, or any other possible contributing factor in a society where women do not have parity in power.

On October 5, 2017, American actor Ashley Judd accused movie mogul Harvey Weinstein of sexual harassment in a story broken by the *New York Times*. A week later, producer Isa Hackett made similar claims against the head of Amazon studios, Roy Price. The accusations were received publicly, with the false-until-proven-true skepticism I describe above, and many other actors and celebrities came to the defense of the character of both men. There was also online and cultural backlash against both women for all kinds of stated reasons, like not reporting the events sooner, not addressing the matter in the moment, and enticing the men into the acts, with that same familiar narrative that men cannot be expected to respect the bodily autonomy of women (Hawbaker 2019). On October 15, 2017, in an apparent effort to demonstrate the reach of the problem of sexual harassment and open a discussion about its commonplaceness, American actor Alyssa Milano posted to Twitter a request that "if you've been sexually harassed or assaulted write 'me too' as a reply to this tweet" (Milano 2017). And people did. Participants did not just post "me too," but they also began sharing their narrative accounts, the storytelling, of their experiences with sexual harassment and assault. #MeToo began trending on social media platforms almost immediately. This led to, as I described in my introduction to this chapter, a global movement that eventually accounted for more than 19 million tweets, plus countless Facebook posts, news and magazine articles, newsclips, and water cooler conversations, as well as resignations and public apologies from prominent men in the entertainment industry and political sphere.

#MeToo has its detractors, including those who criticize it as white feminism and neoliberal feminism. First, #MeToo did not suddenly emerge in 2017, but rather was created by activist Tarana Burke back in 2006 in her work with Black girls who were victims of sexual abuse in the American South. Those who were credited with the spread of #MeToo in 2017 were accused of appropriating Burke's movement. Second, with women posting their narratives of sexual harassment and assault, #MeToo was nearly inescapable, pouring into feeds across social media platforms. This led to other women and men leaving social media or scaling back their own social media usage to avoid the "triggering" aspects of these narratives. Third, *white feminism* is a phrase for describing feminism that centers the concerns of relatively affluent white women and neglects the issues faced by poor women and women of color. This was a point of contention with the #MeToo movement, because poor women and women of color may be too vulnerable to share their own stories without the fear of reprisal, a privilege only women secure in their places and spaces can claim. Thus, the movement that claimed inclusivity could not hope to represent the stories of women who might lose their incomes or face additional abuse as reprisal as they came forward. Fourth, #MeToo was also associated with a kind of neoliberal bootstrap narrative. Some women who shared their stories, like Ashley Judd, relayed those stories as a kind of narrative of personal growth or development from the experience, positioning themselves outside their own narratives, as if a woman could better resist sexual harassment and abuse if she grew older, more experienced, or wiser. The result was a series of narratives that could be read as overcoming the negative consequences of abuse through one's own efforts. These particular narratives ignored the institutional and structural inequities that enable abuses by the powerful, instead framing their abuse and assault as a self-help narrative in the vein of neoliberal individualism.

#MeToo as a campaign is not without its detractors.

Critics have called it triggering as victims choose to leave social media rather than be forced to relive their own traumas. It's been called a privileged example of white feminism. And said that it ignores institutional sources of abuse of anyone "othered" making #MeToo just another neoliberal exercise in self-improvement and bootstrap narratives.

Additionally, as many have pointed out, #MeToo is, in many ways, just more of the status quo.

It is victims asking to be believed in a culture that discounts out of hand the lived experiences of women and nonbinary and nonconforming folx.

So, as the argument goes, this represents the status quo, because instead of assigning responsibility to the perpetrators, victims—largely women and nonconforming persons—once again, are bearing the emotional work of proving their own legitimacy.

Here, I respond to why it is an untenable practice to ask, in the case of #MeToo, women to do the emotional labor and cultural work of proving their own right to call out the behavior of others. Also, here I turn the scope outward and argue against this practice of requiring the already historically excluded to prove legitimacy in the colonial-patriarchal power structures, such as with institutional diversity initiatives, with which academic conference attendees would be familiar. It is commonplace for institutions to forward "diversity initiatives." It is also commonplace to recruit minority members as agents of institutional diversity. This often places women and people of color in the position of "educating" their white colleagues about their own issues. It is important work, certainly, but it is also the kind of work that is placed atop other institutional labor, relocating the burden of changing the institutional culture on the marginalized rather than placing the responsibility on the shoulders of people who are privileged by the existing power structure. Additionally, the emotional labor of navigating confrontation or steering through the near-universal defensiveness people feel when told they are privileged is also invisible work. Again, the dominant culture reserves the right to arbitrate what is true, valuable, and credible, and people working in diversity initiatives must overcome these forces. We often term the confrontation with institutional and cultural power that pushes back against change "resistance," as if the structures themselves are static and changing the structures is merely a matter of amassing enough momentum to shift those static objects. But changing structural bias is not so much resistance or friction against static objects; structural bias is not simply a force commensurate with the stubbornness holding back change. Rather, this insistence that women and people of color educate their white colleagues about bias is a structural procedure meant to retain the status quo. It is deliberate and direct action of settler-colonialism, not resistant reaction. It is the plan, not a side effect.

Here in my talk, I am considering and offering for consideration all of these arguments that mired #MeToo in debate about its effectiveness, its inclusivity, its motives, its value, and so on. And I agree with many of these arguments and claims. It is indeed a privilege to speak back to the power that would oppress you, with minimal or manageable damage to your place, space, identity, and livelihood. It is indeed privilege when I speak about my own experience of harassment, due to all the factors in my life that create a personal, career, and economic safety net for me to fall back on should my admissions be leveraged against me. But I disagree that these problems may invalidate the #MeToo movement. Intersectional feminism is a tool to describe and discuss these complexities, and certainly the #MeToo movement needs intersectional feminism, to represent more people—if the goal is to demonstrate the scope of the problem of sexual harassment and assault. But #MeToo is also a movement of narrative, of describing lived experience, and of building on the strength of a digital network of relations. It is storytelling, and Indigenous methodology gives us storytelling as theorizing and as data. In TribalCrit, Tribal critical race theory, the Indigenous branch of critical race theory, stories *are* data, and #MeToo was and is rich storytelling, with data (Brayboy 2005). Indigenous storytelling does not operate with the western skepticism that requires participants to climb beyond some mythical amount of "acceptable" evidence, in the face of the greater cultural power structures of patriarchy and settler colonialism, in order to offer "proof." Decolonizing rape culture would allow for precolonial considerations of gender, not embedded in patriarchal colonialism and settler colonialism, to dismantle some of this intertwined power matrix that suppresses women's experiences with sexual harassment and sexual assault.

> *But I contend here that these "you're doing it wrong because you're not feminist enough, intersectional enough, collectivist enough, careful enough" are an incomplete reading of the rhetorical practice of #MeToo. Because you can also read #MeToo as a rhetoric of survivance of people whose experiences are persistently silenced and erased.*
>
> *#MeToo is the practice of saying:*
> *I survive.*

> *I persist.*
> *I am legitimized in this place and space that is my own when I speak it.*

Here, I am instantiating Indigenous practices for those who are relaying #MeToo stories. I have legitimacy because I am a woman, a storyteller, a Native telling a story, offering a talk, on survivance and precolonial power. I extend this legitimacy to other storytellers experiencing oppression.

> *The second Indigenous practice I want to discuss, I would like to direct at our allies.*
> *When I was asked to give this talk, it was suggested to me that I say something provocative. If you know me, or if you just know me from online.*
> *You know I can be provocative.*

Here, *provocative* is given several meanings, and I paused on this word while giving my talk, to emphasize that I am calling on more than one meaning here—the thick meaning of *provocative* to me. My social media presence, like that of many people, has pictures of my cat and displays of pride in my nieces and nephews, but it also includes a great deal of activism. And given that my primary areas of research deal with the colonization of Indigenous peoples and the abuses against women-identifying people and other marginalized groups, not everything I post is comfortable for all audiences. It can be provocative in that it is a catalyst for that discomfort. I also have a snarky sense of humor that tends to skewer institutions and institutional agents. In other words, my humor can be pointed, provoking an audience response to my satire. Finally, I make reference to the intellectual practice of providing a "provocation" to spark discussion. Some of these definitions of *provocative* may be at cross-purposes with each other—my using the term comes packed with conflicts in meaning and intent. These competing meanings are possible because they are contained in a thicker version of provoking. I can provoke audiences with my humor as much as I can provoke a discussion with my scholarship. I do call out a false gender binary system, while also drawing attention to it. My thick provocation here was to direct the next section of my talk to "cishetmale allies." By *cishetmale*, I mean cisgendered, those whose assigned sex at birth aligns with both their identities and their gender performance. Cisgender can be a problematic term, because it may reinscribe the masculine/feminine binary in the terms of cisgender/transgender in the same way *heterosexual* reinforces the heterosexual/homosexual binary. However, in this case and context, I am being provocative, emphasizing the artificial gender and power binaries that underpin the gender-charged online abuse I would next bring into the discussion. In this move, I am expressing mild frustration, even as I speak of respect and value. With *cishetmale*, I invoke the image of a member of the dominant western culture, and I am signaling that I am going to call that person in.

> So, this next point is a provocation. I mean it with all respect and affection for my allies, but I'm speaking largely to my cishetmale allies, not because I think you can't be harassed and abused either in physical space or digital space.
> Because of course you can and of course you are.
> But as members of, and co-producers and co-constructors of the dominant culture, you are uniquely positioned to help us write a new narrative, one vastly different from the one we have now.

I think most people feel confronted when they are told they have privilege, because the difficulty in privilege is that it is immersive and uncritically accepted as the norm. As I have described, Indigenous practice, as well as feminist practice, is to identify and know your own positionality. This is a *reflective* process, as opposed to a self-centric one, and one that is community-oriented in its nature, in the way that these situations and connections are articulated rather than held to oneself. People belong to an extended network of relations, to other people as well as animals, bodies of water, the earth, the elements, and life in all its forms. Indigeneity is to acknowledge not only a person's own place in the relations network but their accountability to other entities in the network as well. I set this practice in conflict with western neoliberalism that considers it a virtue to be concerned only with one's own situation. Supporting or sustaining anyone else is considered charitable, which is an assessment that someone else has less value in the marketplace and then provisioning some measure of a thing for that instance of inequality. It is not partnering with someone simply because we are all relations within the relations network. Rather, it is assessing worthiness and then begrudgingly doling out resources valued on a relative basis. Within the neoliberal framework, each person's experiences are conceived of as individual ones, and each person's accomplishments are also conceived of as individual. Western neoliberalism says your successes and failures are your own, measured by your viability in the capitalist marketplace. People do not accept collective responsibility for the sick, poor, unhoused, elderly, young, and the environment, because these people and things are not profitable or valuable in the marketplace. Within this marketplace system, people feel confronted when described as privileged, because privilege implies that they did not act in accordance with the rules of the neoliberal economy. They have consumed the belief system that the economy is value-free and blind to inequality. This is of course not the case; however, whole identities are constructed around relative success in the neoliberal market, and challenging those identities destabilizes what some people think they know about themselves, and thus they feel confronted.

Our impulse, locked in this framework, is to think about our own circumstances—the things that detract from our own market viability. Academic salaries are relatively low compared to industry workers with similar levels of education. We are all subject to ageism and the western obsession with its own standards of beauty. One of my cishetmale allies may feel disempowered and not privileged or see themselves through the filters of the neoliberal system of bootstrapping. They worked hard for who they are and what they have, and the system does not position them as responsible for the condition of others. They did not harass anyone and do not want to be held accountable for the behavior of others. It is hard to see how an individual has power and privilege when they are subject to market forces, when the neoliberal model argues that everyone begins at the same point, and people earn their own way through different levels of self-success. There are many points where we may not feel privileged, many circumstances and contexts each of us carry. But indigeneity asks us to think of ourselves as part of a whole and know that there are people in the whole whose positions are different than ours, and whose differences are valued. Decolonialism offers a point of resistance to the narratives and frames of capitalism (Powell et al. 2014). What I am discussing here is the notion of relative power. One system tells us that we are independent entities responsible for only ourselves, while the decolonial process I am referencing here *knows* that we are all connected and are accountable to each other, even for the consequences of things we did not do or cannot control. It is *we are*, rather than *I am*.

> *And in your positions of relative power, I'm telling you, provocatively here, that you can't have it both ways.*
>
> *You can't interpret certain digital cultural practices as only harmless, when abusers and trolls know to employ those exact practices to cause harm.*

At this point in my talk, the room was murmuring, because here is the crux of my argument about gendered online abuse, and the example I will cite in a moment moves this talk from general discussion of sexual harassment to my case study of the pornography deployed to harass women on the feminist website Jezebel.com. Too many woman-identifying and other marginalized people are targets of online abuse, yet many of them are told that this is not abuse because it is only pornography, it is only sexually explicit humor, it is only conventional social practices. Here, I resist these easy and commonplace dismissals by pointing out that if someone's narrative says harm is done, then that voice and that experience are important and those stories are valuable and authentic to them. Yes, there are arguments that pornography is harmless, but pornography becomes a threat of sexual harassment and sexual violence when it is deployed to a woman's social media feed in an effort to shame women, reminding them that they are just bodies controlled by men and that they live under the threat of sexual assault. Women-identifying people are more likely to be raped, and women are more likely to be touched without their permission. Women, more than men, face the threat of revenge porn, sexually explicit images once shared voluntarily in an expectation of privacy in a relationship, a privacy that is broken when a former partner shares those images publicly to both shame a woman and reduce her value by sharing her body with others. And this is why we must be careful in how we imagine digital cultural practices to (not) have impact in the physical world.

Here's an example: the humble yet unsolicited dick pic, the artifacts we who do—and even do not—date men receive in our inboxes in heaps and stacks. Of dicks.
Yes, go ahead and laugh. I'm being silly.

And I was being lighthearted here. And I can be lighthearted, because often nude photos are mutual, or they can be flirtation and enjoyed by the intended audience. Katrin Tiidenberg's analysis of dick pics complicates the notion that the sending of these kinds of images is always bad or always good, that they are always aggressive or always appreciated (Tiidenberg 2013). She points to many examples where the photographs are validating to the sender, and where the receivers take great care in their role as validators. She emphasizes the trust and sense of belonging that arise from consensual sharing of these images. But her ethnographic research also reveals some cases where these pictures are harmful. From one interview, a participant responds, "I find that there are some guys that get off on the fact that you're about to see their cockshot without consent. Seriously. For some guys that seems to be the thing, you don't want to see it but you WILL see it because I'll put it in your inbox, I will force it on you" (Tiidenberg 2017). Tiidenberg reminds us that sometimes forcing a dick pic on a nonconsenting audience is a product of rape culture.

Many of the narratives I read with the MeToo hashtag included stories of the first time the victim was sexually harassed. And for most of those people, the first time they were sexually harassed was when they were little girls, and a grown man flashed his dick at them.

Now this practice of flashing little girls gets waved away in North American culture as simple perversion, the trope of the dirty old man. But I'll contend here that this too is a misreading. Because the flashing of little girls is part and parcel of a broader culture that engages in a race to be the first to take a woman's virginity.

I mentioned rape culture earlier, and I gestured to the power of patriarchal colonialism to set a material and cultural value on women's bodies. Here, at this point in my talk, I introduce the ideas of autonomy and consent. Autonomy is the right of a person to act on their own behalf, to make their own choices, and to govern themselves. I have pointed to a number of systems of power and privilege in this chapter that interfere with the autonomy of persons. Consent is the practice of giving explicit permission. In the paradigm of rape culture, women have neither autonomy nor the right to consent or decline, because their entire personhood that would come with its own autonomy is in question. In American culture, the settler state always retains the power of governance over Indigenous peoples and denies their right to consent. Lands are seized and civil rights restricted, and the North American governments proceed with their ongoing colonization enterprise, because it is profitable for them. The personhood of Indigenous folx is undermined by the settler state, and thus self-determination is one of the agenda items of the Indigenous research agenda (L. Smith 2012). Like women's bodies, land is not autonomous, nor is it valued simply for itself; land has value only in the context of what it is useful for and how much market value it may have. Land is a relation, but land does not have the right to either consent or decline. I do not want to say here that these issues are exactly alike, or that they are as simple as I have written them. I note autonomy and consent here to foreground the notions that these issues are common across the colonized, whether bodies, water, or land, and thus the potential for allyship emerges.

Who got there first? This is always the question.

And so, the flashing of dicks to little girls is rape culture. Flashers know what they're doing, and they know what cultural practice they're engaged in.

It can't be just an unsolicited dick pic if it's also a race to take virginity. If it's also sexual harassment.

It is difficult to work when you are targeted by trolls in a shaming and threatening campaign. In 2014, feminist media site Jezebel became the target of trolls posting pornography and rape GIFs to the comments sections of women authors' posts. Now a member of the Gizmodo Media Group, Jezebel.com at the time belonged to the group of websites under the umbrella of Gawker Media LLC. The Gawker sites often reported stories based on tips from readers. The tips could either be sent to a site's editor directly, or be posted in the comments section. The comments section was set up to allow anonymous posting because, as the Gawker executives stated, they wanted to protect the free speech of their tipsters. The Gawker Media Group's overarching policy was that the comments sections in the websites in their group were to be manually moderated in case a tipster left a tip in the comments. The anonymous nature of the comments section, paired with the manual moderation by the editorial staff, meant that if anonymous posters posted violent pornography to the comments section, there was no platform mechanism to block it. Jezebel's all-female editorial staff was charged daily with looking through the comments section to delete any potentially offensive content. In effect, the Jezebel editorial staff was exposed daily to GIFs of sexual violence in the name of freedom of speech. On August 11, 2014, Jezebel's editorial staff took the extraordinary step of posting an open letter to the internet, "We Have a Rape GIF Problem and Gawker Media Won't Do Anything about It." The letter argued that the Jezebel.com editors had repeatedly complained to the Gawker executives about the trolling of their site and the complex kinds of fatigue they experienced from having to screen the comments constantly for violent pornography. According to the editors, the Gawker executives told them that it was just pornography, and there would be no additional tools for moderating the comments because freedom of speech and anonymity were corporate values. Here "trolls were empowered to post violent pornography in the interests of free speech, while the freedom to network and discuss feminist interests were threatened" by those same trolls (Tekobbe 2015).

> I'll give you another example. The practice of responding to women's posts with porn GIFs is rather commonplace, but a really well-known instance was in the summer of 2014 when the feminist website Jezebel was literally under siege by trolls and their sock puppets posting violent porn to the comment threads.
> The Jezebel staff reported the barrage of porn to their largely cishetmale executive editorial team and tech support, who responded that it was only porn. Ignore it.
> Perhaps it was only porn.
> But with porn, as with most things, the context is everything.
> And if it is only porn and porn is harmless, then why do trolls employ porn to shame, humiliate, harass, and threaten women in their own spaces?

I do not mean here that a male flashing his penis to a little girl is exactly the same as males sending unsolicited and nonconsenting penis images or is exactly the same as posting rape GIFs to a feminist-oriented website. I do argue that they are products of the same rape culture I described earlier. The commonality here is in the lack of right to consent or decline. When violent sexual imagery is deployed to a gathering place of women (the comments section), without their consent or mutual participation, this is sexual harassment by definition because of the context. The primarily female, feminist audience was using the space to discuss news stories and opinion pieces not inviting engagement with violent pornography. When this overtly threatening material is deployed to a workplace of women, this is the most commonly discussed form of sexual harassment—workplace sexual harassment. When someone flashes a little girl, he is generally triggering shame and feelings of threat. And according to the laws that protect them, little girls cannot give sexual consent in the first place. These instances are not the same, but they come from the same power matrix of patriarchal colonialism.

> *So, I'm saying my cisthetmale allies can't have it both ways. When you listen to our stories, you can't just brush them off as harmless if we say these acts are causing us harm. Because it can't be both a harmless dick pic and sexual harassment. It can't be harmless porn and sexual harassment. Not when it is embedded in the context of our lived experiences.*
>
> *These cultural practices of dick pics and porn can have specific harmful meaning to victims, a meaning that can be different than the dominant narrative.*
>
> *And dear allies, we need you to hear the difference, to be aware of the cultural practices, and to understand that your narrative isn't the only narrative, or even the most important narrative in a given context.*

Here, I contend that the Indigenous practice of telling stories, of speaking and listening, of co-creating knowledge from these stories is a valuable tool for breaking the gridlock of the cultural skepticism and dismissiveness that surrounds women's issues. With Indigenous rhetorical thickness, there is open space for accounting for the emotional and spiritual fatigue of dealing with rape culture that the editors of Jezebel report, as well as the emotional experiences of participants of the #MeToo movement. In fact, one of the obstacles women face when telling their stories of sexual harassment and assault is that their tellings are emotional, not rational, and only rational, evidence-based "claims" are given space in western culture—with *rational* being a problematic frame defined by western values and, yes, feelings. I contend here that woman-identifying persons who stand at the intersection of neoliberalism, colonialism, patriarchy, capitalism, race, class, and gender have stories that are valuable data and are critical ways of knowing in the world. Social media networks people in a myriad of ways, and our stories spread across those networks, across relations. And we can speak these stories into the network because we have that role in a community. We are not silenced ones but egalitarian speakers. Our stories are knowing.

> Listen to women and nonconforming folx. We're telling you something.
> We're saying
> We are surviving this
> We persist.
> We are legitimized in this place and space that is our own when we speak
> And we need you to listen.
> I thank you for your attention and your time.
> Yakoke.

This chapter is about being Indigenous in a settler-colonized world. It is about being woman-identifying in a patriarchal world. It is about being marginalized by a homogenizing, hegemonic narrative of whiteness and maleness, in physical and online spaces. It is about being a digital activist in a neoliberal technocracy. I argued here that cultural, Indigenous, and intersectional frames were necessary in this example of internet research, because they have the potential to disrupt the narrative that constructs the typical online user as "white male." I argued that neoliberalism is both a self-replicating power system and one that appropriates every identity and experience it touches, subsuming them into the capitalist system. I argued that Indigenous notions of good relations as a research positionality are one possible way to disrupt neoliberal systems. And, then I demonstrated the need for Indigenous, intersectional, and cultural methodologies, using my two case studies of #MeToo and Jezebel.

During the question-and-answer portion of the plenary session where this talk was originally given, I was asked if I thought digital activism had any hope of change-making, or does it simply reproduce existing power structures. I remember responding that I did not know, but seeing the session tweets, the online joining of my conversation by attendees, was making me feel hopeful. In general, neoliberalism is neocolonialism, with wealthy corporations serving as the new imperialists. And while this neocolonialism most obviously and asymmetrically impacts the historically excluded, its effects are felt broadly. For example, Colin Kaepernick, the American football player, began kneeling during National Football League games through the performance of the American national anthem to protest police brutality against African Americans. Not long after, the shoe and sportswear giant Nike produced an ad campaign appropriating the protest to sell its goods. The "believe in something even if it means sacrificing everything" campaign raised Nike's profits by 5 percent (Abad-Santos 2018). Here, with Kaepernick, a poignant social justice protest is subsumed into the system to produce corporate profits. Which leads to a question of survivance that moves beyond Indigenous peoples. How will we retain essential parts of our identities and communities in the face of neoliberalism? Here I argue that Indigenous peoples have stood in resistance to colonialism for more than five centuries—we survive, we persist, this place is ours because we tell stories of ourselves into being, both in modern practice and with the voices of our ancestors. And we have accomplished this resistance through the sustenance of the networks of our relations. I invite you to consider this survivance and perhaps extend these notions to your own path of resistance.

3
Skoden

Indigenous Identity Construction through Facebook Memes

Why the Rabbit Has a Short Tail
A very long time ago, only the Red People and the wildlife were on this land. At that time, the rabbit had a long tail.

Early one very cold morning, he was hopping and playing. He looked toward the trail and saw a fox coming. The fox had some fish. "Wow! I'll ask him where he caught the fish," thought the rabbit. When the fox arrived, Rabbit asked, "Fox, where did you catch those fish?" The fox said to him, "I caught them at the branch. Although the branch was frozen, I dug a small hole in the ice and put my tail through the hole. I sat there for quite a while and my tail began to get heavy. I pulled my tail out and the fish were hanging on it."

The rabbit hopped very quickly toward the branch. When he got there, he dug a small hole in the ice and put his tail through it. It was very cold but the rabbit kept sitting on the ice. When he thought he had enough fish, he pulled his tail but it was frozen to the ice. He couldn't take his tail out so he pulled again. He pulled so hard that his tail snapped. That is why the rabbit has a short tail. (Niottak Hullo Chito 2019)

https://doi.org/10.7330/9781646426478.c003

Voter Suppression Is a Feature of Settler Democracy

During the run-up to the November 2018 midterm elections, a news story broke about voter suppression of Native Americans in North Dakota. On North Dakota reservations, like many reservations across the United States, many residents are given only PO boxes due to the sprawling nature of their lands. Therefore, they necessarily use those PO boxes as their addresses on official government documents. Other tribal members might not use documentation issued by federal and state agencies; they may use only tribal identification cards that do not list a physical address. In other words, the federal government allotted remote lands to people, lands that were too remote to easily access, so PO boxes were implemented as a government solution. Then, the government created new voting rules to require a physical address on state identification to validate voter identity (Reilly 2018). The tribes sued and the Supreme Court ruled in favor of North Dakota in requiring a physical address to vote. Now, PO boxes are invalidated as official addresses. The result here is that Indigenous folx are disenfranchised by a settler-colonial government in a system set up by and perpetuated by that same government (Booker 2020).

Voter suppression is nothing new in Indian Country. The Indian Citizenship Act was passed in 1924, yet on some reservations only in recent decades has there been access to voting. In other places, ballots are collected and brought to polling places. In Montana, where there are twelve Native American nations, it is illegal for organizers to use ballot collection services to collect ballots from rural areas (Friel 2022). The same holds for Arizona, home to twenty-two federally recognized tribes, many living in rural and remote areas, as well as at least fourteen other states. There are other complications as well. With the passage of voter ID laws, it can be difficult for Native Americans to meet the documentation requirements to obtain the ID cards, given that birth certificates may not be recorded, or may be recorded in different names, and other non–photo ID documents, like bank statements and vehicle registrations, may not be relevant to the Native voter trying to obtain an ID card (Dunphy 2019).

In fact, the story of 175 years of BIA (Bureau of Indian Affairs) oversight sediments in a narrative of lost lands and assets, poverty, a fragile social safety network, and geographical and technological barriers to Indigenous peoples engaging as full citizens. The prison industrial complex plays its role as well given that Indigenous peoples are incarcerated at a rate 38 percent higher than the national average, a number that exceeds other demographic groups. Native Americans are also overrepresented in unhoused populations (Domonoske 2018). Indigenous people's lands are encroached on by mining operations, oil pipeline construction, and toxic waste disposal. The complications of federal jurisdiction on Native lands have contributed to the rates of missing and murdered Indigenous women. And finally, Indigenous peoples are subject to gross social biases, such as stereotypes of drunkenness, laziness, and implications of being less-than-human, left over from the colonialisms of "savagery" and non-Christian spiritual practices. Together, these laws and practices disproportionately affect Indigenous populations and communities and work together to actively suppress the Native American vote. I will stop here, as I am sure you understand that Indigenous peoples' right to autonomy and self-determination continues to be denied by the settler-state machines that are alive and well today. I am sure readers understand that political identity and activism of Native Americans are suppressed or made invisible by everything from deliberate erasure to general public disinterest.

Being Indian, Doing Digital

This context gives rise to questions of how Indigenous folx resist the silencing, engage current issues, build individual and collective digital identities. It asks how social media can be used as a nontraditional way of accessing power, as marginalized identities' and groups' means of social support, and as the self-presentation of marginalized groups in online spaces. The first question I want to answer, however, is why do this research at all? Why study Indigenous digital identity construction? Why do I think audiences will care about the digital identity work of 2 to 3 percent of the American population? And why study it in the form of memes? I have two primary reasons for exploring memes, but given the widespread use of memes and the proliferation of meme-sharing groups on social media, I think memes could use a lot more research and discussion. My first reason is that while digital identity construction is a common research area in internet studies, little research has been done in Indigenous digital practices. There are a number of texts looking at intersectionality in digital spaces, including *Intersectional Internet*, edited by Safiya Noble and Brendesha Tynes (2016), and *Digitizing Race* by Lisa Nakamura (2007). These texts do much to theorize the internet as an intersectional space where actors and activists in various race and gender studies enact change. Very little, however, has been written about digital forms of composition and identity in relation to indigeneity (Haas 2007; Tekobbe

2019). As I wrote in chapter 1 of this text, I think Indigenous digital identity construction is unique because its thickness—its layered collection of stories, with its range of details and storytellers, held together by land and history—is unique. I am not only saying that traditional Indigenous storytelling is unique but also that unique storytelling is translatable to a digital platform.

My second reason to research memes is to add my theorizing to a growing body of work about internet memes (Downes 1999; Knobel and Lankshear 2007; Shifman 2013; Sparby 2017). Memes are important internet media that are used for a variety of purposes, like reacting to posts, commiserating with a contact, or sharing humor. There are websites devoted to producing and sharing new memes and promoting existing memes. "Internet Meme Database: Know Your Meme" is one such website that acts as an internet resource to learn the history of a meme and interpret its meaning and known uses. The primary marker of memes is that they are "contagious patterns of 'cultural information' that get passed from mind to mind and directly generate and shape mindsets and significant forms of behaviour and actions of a social group. They include things like catch-phrases, clothing fashions, architectural styles, ways of doing things, icons, jingles, musical riffs and licks, and the like" (Knobel and Lankshear 2007). In other words, internet memes, also known as online memes, are memes that are produced and reproduced and spread across the internet through messaging apps and social media platforms (Shifman 2013). Knobel and Lankshear, and also Shifman, write that a fundamental factor of an internet meme is its "intertextuality," meaning that "memes often relate to each other in complex, creative, and surprising ways" (Shifman 2013). Memes are referential, often mashing up and remixing cultural tropes for new audiences, or the same audiences with new material.

Thinking about Digital Decolonizing, or Theorizing Unsettling with Digital Technologies

My study is a visual-rhetorical analysis of internet memes that are created by administrators and moderators of a Facebook group that post and share socialist-leaning political viewpoints by people who self-identify a kinship to American Indian reservations and communities. Shifman (2013) argues that memes are an example of identity and affinity constructions in digital participatory culture. In other words, one of the reasons people online build and circulate memes is to aid in online group identity construction. In addition to social media and meme research texts, I am also using research lenses specifically from Indigenous methodologies and critical race studies. Shawn Wilson's work on Indigenous research methodology is critical here in that rather than the western approach of locating knowledge in the intellect and as objective, Indigenous peoples approach knowledges through the personal—through senses, intuitions, emotions, and feelings (2008). I identify the influences of indigeneity and listening in my methodology and analysis.

I begin with three research questions:

How is the meme an effective way of constructing political identity and kinship?
How does this specific Facebook page access political power in this context?
How do this page and its community use social media for social support of Indigenous political identities?

Ethical Considerations Online, Some Choices

When researching online, it is important to consider the ubiquity of algorithms and other platform affordances that both lend themselves to making information more widely searchable and make it difficult to draw a line between a person's public and private internet usage. Here, I am researching a Facebook page, and I have taken a number of precautions to help shore up that public and private division. By "public Facebook page," I mean a page that anyone can like or follow. In this case, although the page has not been updated in a few years, it is still active and people like and follow it as well as comment on its images. When I began this study, I was a bit concerned about its public nature as well as the rawness of its content. Native humor can be black humor, because the issues that are being mocked are not traditionally thought of as amusing, such as the genocide of peoples, the kidnapping of children, the spread of disease, and the corruption of governments and individuals. Poking fun at this painful history can lead to even more difficult satire. So, I wanted to be careful.

In brief, I have made several ethical choices with my meme research in a good-faith effort to prevent its followers from being identified. All of their Facebook posts are public, but research ethics requires me to consider how I might avoid encouraging harassment of the posters and commenters. I retain all the images, unaltered, in a digital file on my home server. In my research, I have blacked out the names of posters and commenters.

Digital Methodologies

First, in my selected Facebook page, Rezzy Red Proletariat Memes, I screen-captured all of the memes posted between October 1 and November 2, 2018. I chose October 1 as the starting point, because that date was when I first explored the memes of the page in relation to voter suppression in the 2018 midterm elections. After I collected the memes, I coded them as general political, Trump-specific, Department of Interior–specific, Immigration and Border Control–specific, responses to white supremacy, specific negative responses from the public to the page, and other memes. I also coded them by the emotion and the historical narrative they seemed to invoke. I then charted these codes and made some observations about what types of media were shared, including which codes had the most responses and whether types of memes were collected on specific dates or were spread evenly over the date range.

For my next steps, I visually analyzed all the memes that are Trump-, political-, and white supremacist–specific. I chose these because they seem most closely related to the page's identity as a group for socialist-leaning Indigenous people resisting the status quo. In this analysis, I looked for common themes and Native American–specific themes to draw some conclusions about how Indigenous resistance as a political identity is constructed and where, specifically, emotion and motivation are invoked as part of the Indigenous approach to research as ceremony. Through this analysis, I discuss and respond to my research questions.

One potential problem with my methods is that I am relying on the Facebook algorithm to show me all the memes in my date range, but I cannot control how the Facebook programming applied to the raw data of memes on the timeline is represented, so I may be missing memes. I am not analyzing these from an idealized sense of neutrality; rather, I am viewing these as a Native American with shared identity practices and political positions, which I hope improves the project because I can note in-identity meanings and invoke less well-known histories.

My methodologies are feminist, Indigenous, ethnographic, and rhetorical-analytic. Because feminist practices are transparent, culturally situated, and participant-foregrounding, I am writing about my work and my relationship to it. As I have explained, Indigenous knowledge practices are different from western knowledge practices in that western knowledge is centered in the mind—a "rational" process of a scholar in a room with their books and their data. Indigenous knowledge practices center relationships. Indigenous knowledge is co-constructed between people, as well as being experiential and imbued with emotion and intuition. This Facebook page represents a space that is co-constructed by the actors thinking about and commenting on political and cultural identity issues.

Other slides critique white feminism's lack of intersectionality, Elizabeth Warren for not using her power to foreground Indigenous political issues, a federal response of police action to instances of Indigenous resistance, and Indigenous moderates who prioritize their own power and comfort over advocating for their people. I contend here that the satirical remix of Indigenous culture, popular culture, and American politics helps create a subversive Indigenous identity and acts as a means of community-building.

Applying an Indigenous Eye to Data

My first question, How is the meme an effective way of constructing political identity and kinship?, accepts and builds on current research showing that the sharing of internet memes operates in affinity spaces to build group cohesion and identity (Knobel and Lankshear 2007). I look at the subject the meme relates to and how I know this information. Often, I know the subject of the meme because the picture makes it obvious, such as the meme "Obama and Trump Indian Policy," which features a photograph of militarized police facing off against unarmed Native protestors at Standing Rock. I am familiar with images of militarized police, and I think anyone evaluating this meme could determine this subject; the context, however, comes largely from knowledge of Native issues. A Native reader would likely know about the significance of the conflict over clean water at Standing Rock. They would likely know about the pan-tribal group of Native Americans and their allies who camped out together in protest of the "black snake," the Dakota Access Pipeline, and how it might cross sacred lands and waterways. That the media reported the camp was comprised of unarmed, nonviolent campers makes the story of a militarized police response starker, certainly, but the image is made even more poignant by the cultural memory of the Indian Removal Act, where tribal peoples were relocated at gunpoint by the United States military. The memes that I studied, a sampling of which I discuss here, largely required contextual knowledge to understand the meme and why it was shared with the group. And this need for contextual knowledge, when posting the meme to a Facebook group, suggests that the people in the group share the contextual knowledge(s). In short, I know it is an Indigenous page, not only because the page name alludes to Native issues but also because the memes posted required shared contextual details, details that would be known to those engaged in Native issues, news, conflicts, and protests.

My second question, How does this specific Facebook page access political power in this context, originally referred to the physical contextual details of the memes. When I wrote this second question, I was thinking about geography and land, perhaps even mapping the physical contexts of the memes, as some were concerns that were generally important to all Natives even though they might have originated from a specific Nation, even while they were widely relevant to Native peoples. For example, the images of Pocahontas are significant to many Natives, because her story is not the Disneyfied cartoon version but a story of settler colonialism and the rape and subjugation of Native women. Even as she can be seen as a universal figure, Pocahontas might be more significant to the Anishinabeg, who are relatives of her Powhatan peoples who once lived in the Chesapeake Bay region, and I might pin that region on a map. I might pin Massachusetts, the state that Senator Warren, the subject of the Pocahontas meme, represents. And, I may have pinned Washington, DC, where the Senate meets and where the White House that Trump coveted is located. Land and our ties to it are deeply significant to Native peoples and identities, and so I was thinking about places and spaces when I started this project. The places represented stood out to me because of how I physically locate Native folks and their land. However, when I started reading the memes deeply, I realized that there was more going on than a space- and place-based context, for example, Standing Rock. I started thinking in terms of the knowings and truths that made up the context of the meme.

In thinking about contextual knowings and truths, I return to Pocahontas specifically because she is widely recognized as a Disney princess. This is a western knowing of Pocahontas, the subject of the 1995 animated feature film that, according to IMDb, grossed $346,079,773 worldwide and won two Oscar awards for its soundtrack. Additionally, a western knowing of Pocahontas is that she rescued her love interest, John Smith, during a conflict between her people and the settlers at Jamestown. She is a hero compared to the "savages" she came from, mending fences and saving lives of white folks. An Indigenous knowing of Pocahontas is more complex than the Hollywood version. Indigenous Pocahontas is a source of angst and anxiety. She is an uncomfortable figure for the way she has been co-opted by western history and culture. She may have been "married" to her white husband(s), but she was probably a victim of trafficking and likely a rape victim. Pocahontas stands in for all sexualized Native women. She is a settler-colonial fantasy of the attainable exotic through Manifest Destiny—as settlers felt entitled to anything or anyone they could see, take, and claim. Pocahontas is an uncomfortable figure because although she is perhaps a hero to young women, she is also another symbol of the violence of settler colonialism against Native peoples. Therefore, Pocahontas contains western cultural knowings, Indigenous knowings, western historical knowings, and Native truths. This is a shared construction of identity and knowledge construction that contextualizes the meme for the Facebook page and its audience. I will discuss Pocahontas in more detail a bit later in this chapter.

My third question, How does this page and its community use social media for social support of Indigenous political identities?, is the question I began with, but not precisely the one I ended with. As I worked with these memes and read through the comments made about them, I began to think more closely about what I meant by "Indigenous political identities," and I realized that in terms of political identities, I was talking more about political knowings arising from Indigenous worldviews. For example, while there are more than 300 million Americans, we can imagine a kind of unified American worldview that espouses personal liberties, private property ownership, apparent monotheism, and free-market commerce. Or as I have discussed within this text, we can imagine a system of colonial capitalism originated by white religious settlers. These white religious settlers brought with them a distinct notion of individualism that over time was codified into America's founding documents (codified along with sexism, racism, and genocide). Thinking along these lines, I describe the difference between Indigenous worldviews and western worldviews in that Indigenous worldviews are a fusion of land (place and space), relatives, knowings (histories and stories), and bodies. Or, as John A. Grim writes in "Indigenous Lifeways and Knowing the World," indigeneity is an "organic relationality of lifeway, land, and Indigenous knowledge as mutually interactive process" (2009). Or, as Kristin Arola (2017) writes, we can describe an Indigenous identity as *doing* Indigenous rather than being Indigenous. An American identity, for example, is a collection of individualist-focused ideologies, and Indigenous identity is a lifeway with people living in and constructing its network. So, the question about Indigenous political identities transforms into a question about a way of acting in a world that protects this lifeway—acting from a collective, collaborative, and communal sense.

Arola also writes, from her chapter "Indigenous Interfaces" in *Social Writing / Social Media: Publics, Presentations, and Pedagogies* (Walls and Vie 2017), that she has a hard time thinking in the context of self-interest, or what is best for herself, as an actor in a departmental service assignment. I describe a similar situation in graduate school in the introduction of this book, in that I cannot simply do what is best for myself without considering the relational network to which I belong, particularly the opinions of the women I consider wiser and elder to me, my late grandmother, my aunt, and my mother. In the intervening years since I began researching and writing this book, I have added my sister to this list, not because she is significantly older than I am (she is not) but because I have grown into the role of auntie to her children, and I prioritize my advisory place in their lives in a way I could not when they were very small. I realize, writing this, that I do not know very well how people who descend only from settlers, or engage life only from settler frameworks, make political decisions or choose political positions, because I will always choose mine for what I have been taught is the best for everyone. I will always choose mine to sustain a particular lifeway. For example, Indigenous folx want their lands back, because those lands are sacred but also because we believe ourselves to be stewards of the land, and we do not typically exploit its natural resources. There are exceptions to this example. There have been a few industrial decisions made on tribal lands in cooperation with settler industries under the advisement of environmental protection reporting and analysis. But typically, like at Standing Rock, the Water Protectors were concerned with the safety of the Mississippi and Missouri Rivers. Yes, the Protectors had environmental aims: protecting the drinking water for the surrounding communities, then protecting sacred sites and burial grounds. But Water is a relation, and protecting water in the network of relations is protecting the network itself. Water and people are one. The collective political identities of Indigenous people extend to the water. Water is Life.

Political identity is a collective identity for Indigenous folx, and so my analysis of the findings of this study of memes through my research questions becomes a lens of looking for and at what is being protected by communal political action. What is being threatened? What is being protected? What communal epistemology is being undermined? These are my questions now.

Fucking Stoodis

The first meme I will discuss is an image of a Native American male in traditional clothing and an eagle feather headdress. He is posed on top of a bluff with the bow drawn. The text reads "*when you see the 'invisible hand'

FIGURE 3.1. "When You See the Invisible Hand" meme.

on its way to commodify your land*." This post received 118 likes and was shared 38 times. It had two comments—one was a series of laughter emojis and the other was a Facebook username tag. Placing asterisks around text on Facebook signals to the audience that this is a thought, or a setting, not dialogue. In this case, it represents setting a scene, one of a Native person observing in the distance the settlers coming to exploit the land in some way, either through the redistribution of Native lands or the extraction colonialism of removing resources from the land. The Indigenous-specific slang of "stoodis" means, in an Indigenous and American Indian reservation context, "let's do this." The phrase, as I have mentioned, can be traced to American Indian reservation slang, and that is important for at least two reasons. The first is that reservation lands are collectively owned by the people of the reservation, and so the exploitation of land is the exploitation of everyone. The second reason is that this slang is commonly known by many Indigenous peoples without respect to specific tribes or reservations. It stands in as collective identity. So, with the communally held land comes the collective defense of the land in the phrase "let's do this." Or let us do this. Us.

The text of the meme introduces the critique of capitalism by reference to the "invisible hand," and settler colonialism by reference to the commodification of historically Indigenous lands. The invisible hand refers to the philosophy of Adam Smith, who in the eighteenth century introduced the idea that people acting in their own self-interest could bring about unintentional benefits for all. It is a phrase learned by students in economics and civics classes all over the United States and is synonymous with the supposed moral good of capitalism. Imposed over the image of the Indigenous warrior drawing a bow is the text "fucking stoodis," which signals a call of resistance to capitalism and settler colonialism in defense of the land and, thus, the network. If I think about my research questions, the question of whether the meme is effective at creating kinship is clear, because the meme is grounded in Native kinship. In terms of the second question, the collective knowings here are also clear—that Native people and the land are inseparable, and the people protect the land. Finally, in terms of what is threatened or protected in terms of lifeways, the land is threatened, but the people are threatened too because capitalism would, if allowed, usurp their collective identity for individualism and self-interest. This meme was the first I analyzed, and from it I moved to interrogate more on the page.

Pocahontas

This meme, liked 192 times and shared 71 times, is in the internet convention of three vertically stacked images with a caption next to each image. This three-image format in memes is used to demonstrate progression between the top meme, through the center meme, to the bottom meme. In the Pocahontas meme, the top image is a historical engraving of Pocahontas by Simon de Passe in 1616. Pocahontas is in European court dress, and her face

FIGURE 3.2. "Netflix Adaptation" meme.

is set in a serious expression. The text next to the image reads "Manga." The middle image is of Pocahontas from the titular Disney film, with flowing hair and a distant, romantic look on her face. The caption next to the image reads "Anime." The final image in the progression is of Elizabeth Warren. The caption next to her image reads "Netflix Adaptation."

This is an interesting meme to me, because it is both a response to President Trump's derision of Senator Elizabeth Warren when he refers to her as "Pocahontas," and it is a response to Senator Warren herself. At points in

Warren's past, she has identified as Native American because she had been told, according to her family lore, that distant Native American relatives are part of her genealogy. Trump mockingly calls Warren "Pocahontas" in order to draw attention to what he claims is a fake identity used to gain advantage in her earlier life. The progression starts from a posed engraving of Pocahontas in Elizabethan dress that hides her probably long hair and likely hides her tattooing, a real-life Pocahontas, but rendered for a seventeenth-century colonial audience. The animated version of Pocahontas rendered by Disney is also removed from a real-life woman, even more so, because this version is a cartoon and drawn to be palatable to a twentieth-century audience, a beautiful Indian princess with no tattoos in sight. Elizabeth Warren as the "Netflix adaptation" is even further removed from the original. This is a pop-culture reference to the commercialization and commodification of made-for-TV adaptations that gloss over historical details in the name of entertainment and the broadest possible audience appeal. None, however, are representations of the original woman; all are colonial reinterpretations. The moderators of the "Rezzy Red Proletariat Memes" group post memes that reject both President Trump and conservative politics. This meme does not specifically call out Trump, but I argue it exists because Trump raised the specter of Warren as an Indian princess in the first place. The meme, and those who liked it, also reject Warren's shallow identification as Native.

In 2018, Warren was known as the senator from Massachusetts, but she was originally born in Oklahoma, a state with thirty-nine sovereign Indigenous tribes. My own family are Choctaw, originally relocated from the American southeast (Mississippi and Alabama) to reservation lands in Oklahoma. I imagine, given the context of so many federally recognized tribes calling Oklahoma their home, that it is likely common for many Oklahoma families to have stories of Native ancestors. In fact, Warren reported that her knowledge of her Native ancestry was anecdotal from family stories. When repeatedly ridiculed by Trump for her reference to her Native ancestry, Warren further antagonized members of Native American tribes by taking a DNA test demonstrating that while she is mostly of European ancestry, there is evidence that she is of Native ancestry from at least ten generations ago. This DNA test was problematic for many reasons, but primarily because DNA "evidence" is fraught, and Native American tribes do not use DNA evidence to determine their own membership. Most use some form of demonstration of lineage, for example, "blood quantum" evidence, or evidence of direct descendance from an original name on the Dawes Roll. As I explained earlier, many view the use of blood quantum as a tool of genocide. But this is not the primary complaint of those Indigenous peoples angered over Warren's "evidence" of a DNA test. It was because she does not live an Indian lifeway, engage in a network of Native community members, or represent Native concerns in her Senate work. Warren, as it goes, is just another, even further removed, imposter.

As it was in 2016 during the presidential race, so it is today: it is not uncommon for people to ask me for my opinion on Elizabeth Warren, as if I were a kind of expert speaking for all Indigenous peoples. As I have said before, I am not. I only speak for myself and as a child of my family. However, when I identify myself as a Native American, it is because I am a direct descendant of the Dawes Roll, and even more importantly, I have a relationship with my tribe. I engage in Indigenous issues, and I was raised as part of an Indigenous network. In fact, some of the Choctaw language in this book was translated for me by members of the tribe who speak the language, because that's what family does for each other. So, when I am asked about Warren, I remind the asker that the use of *Pocahontas* as a slur is racist. The appropriation of Pocahontas into a western narrative is racist. Warren, in my thoughts, is ignorant of tribal issues and customs. I do not know what she personally feels about Indigenous people, but Indian Country expects more of her than identifying as Native for a job because of distant family lore. You are, after all, the people who claim you, and she is not as yet claimed by any peoples. That is my thinking at this time in my life. Perhaps it will change as I grow older and wiser.

Here's How America Uses Its Land

On July 31, 2018, the financial site *Bloomberg* published "Here's How America Uses Its Land," an analysis and complex graphical representation of land usage in America (Merrill and Leatherby 2018). The article is interactive, and users can click though various data visualization charts of America's agriculture, national parks, urban areas, rural areas, federal grazing lands, and more. The final image is an outline of the United States with proportional areas blocked off for each one of these individual categories.

The Rezzy Red Proletariat Memes group repurposed this outline, removing all the captions and replacing them with variations of "stolen land," like "stolen west side" for the West Coast, "bottom of stolen land" over Texas, "stolen east side" over the Carolinas, and so forth. The message here is, of course, that the entire colonial enterprise of the US government and its settlers resides on stolen Indigenous lands. Efforts to overwrite this history with data visualizations

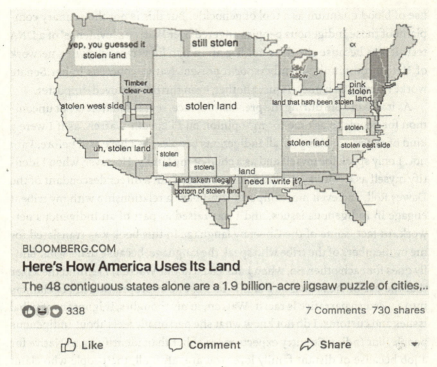

FIGURE 3.3. "Here's How America Uses Its Land" meme.

are corrected. This meme was responded to with hearts, likes, and laughter 338 times and shared 730 times. The nine comments on the meme are mixed, with most affirming that Indigenous land was stolen, and one person dissenting and claiming that ancestral land is an outdated idea. A comment has been deleted, but a response to it still exists. The response to the deleted comment is "when you buy stolen goods, that doesn't magically un-steal them," which likely refers to some of the treaties with Native peoples—treaties that today are viewed with a great degree of skepticism, because they were either not honored or Natives were vastly undercompensated for the land by means of coercion.

As I reflected on my research questions here, I did not have to dig very far to get at the gist of this meme. The title "Here's How America Uses Its Land" tells us about the origin of the infographic, namely that the writer of the article did not think about what the land was before it was colonized and then "used." Land is its own entity, and we Indigenous folx are in a relationship with the land to protect and care for each other. The American efforts to divide the land into different owners and uses is an overt colonial move. The truth of the meme then is resonant—this land is stolen land. The worldview of the meme is a repudiation of settler colonialism that conceptualizes land as coming into existence when it has an individual owner and a practical use, a point of extraction. This is settler-colonial capitalism. What is threatened here is everything, because the people and the land are one, and the theft of the land is a theft of life. Not a way of life, but a lifeway—a mutually interactive process of being. When the land is stolen, the people cannot protect it. This is the source of the #LandBack movement of Indigenous peoples demanding to have lands returned to them. The settlers conceptualize this return as an exchange of ownership, a reparation where one makes a settlement approximate to what was taken. But reparations are not the meaning of #LandBack. The land must be returned because it is people and life.

Obama and Trump Indian Policy

Titled "Obama and Trump Indian Policy," this meme, with 68 reactions and 160 shares, is a four-image grid designed to compare the images on the top with the images on the bottom. In the upper left corner is a smiling President Obama giving a thumbs-up sign. This image is captioned "Obama's Indian Policy." The image to his right is a phalanx of militarized police and their paramilitary vehicles at the protest at Standing Rock, at the site of the Dakota Access Pipeline (DAPL). This image is captioned "Lots of Cops." The bottom-left picture is a smiling President Trump giving a thumbs-up sign. The image to his right is the same phalanx of militarized police and their vehicles, with the caption "More Cops." I note here that four people disliked this post, and it is the only post with dislikes on it that I have shared in this chapter.

I have already discussed Standing Rock to an extent, in terms of what the protestors were defending, as well as their lifeway and worldview: Water is Life. I have discussed the role of the Water Protectors defending the water and why their epistemologies support this. So why do settlers send militarized forces in response to Indigenous protest? Indigenous folx characterize their protests and advocacy work as protecting the land and the water, raising awareness of settler violence against women and children, seeking cleanup of toxic environments left over from nuclear testing and mining, and more. Once you conceptualize that "protests" are people speaking for land, water, murdered women and children, and other relations who cannot speak for themselves, you can disrupt any settler lenses you use to evaluate crowds moving together with signs, regalia and ceremonial objects, T-shirts and other

FIGURE 3.4. "Obama and Trump" meme.

forms of shared identity, and the aims of these groups come into focus. Water Protectors protect water that cannot speak for itself. Settlers, however, "see" the protests of Indigenous folx as agitation for Indigenous interests at the expense of settler interests. Settlers bring "us against them" into a situation where there is resistance to settler motives. The settler key concepts here are agitation and zero-sum competition. Settlers send cops to deal with BIPOC who are seen as "agitated" and a threat to settler property and agendas. Settlers send cops to maintain control of assets. Indigenous folx bring their voices to protect relations. These are very different conceptualizations of land and people. I challenge you to decolonize your views of political and social protests, to puncture the easy paradigm of a "reasonable" state versus "unreasonable" ("savage") mobs. The settler state does not need anyone to defend it—it wields its control and power freely. Instead, listen to who is speaking and who and what they are speaking for.

The Indigenous peoples' lifeways and worldviews that I have described are beliefs that the protected land is an environmental boon for all of us. That the purity of the water is beneficial to all of us. Yes, Indigenous people are close to the land—they are relatives. But settlers are doing themselves harm by polluting the land and the water. By stripping the land of its resources and damaging its ecologies, settlers are harming themselves. Water is life. Land is a relation. And all people are included in this lifeway.

*I Promise I Won't Get All Political *Three Minutes Later**

"'Three Drinks Later' is an image macro series in which photographs of well-known extremists and radicals are juxtaposed beneath the caption 'me: i promise i won't get all political / 3 drinks later' as a way to poke fun at one's tendency to engage in emotionally charged political debate when inebriated," according to the website *Know Your Meme*. This meme dates to 2016 and has seen widespread reinterpretation and repurposing across social media.

In this example of the popular meme, several Natives are sitting on the ground, in a circle, in conversation. One appears to be eating while others are gesturing with their hands and exchanging eye contact. At least five more Natives are approaching the circle, some on horseback. The scene is recognizable as a social gathering because of the talking, eating, and gesturing. (Other than the Native dress, this is what a party I would throw in graduate school would look like. These days, I try to have enough chairs for everyone in a nod to accessibility, but as a graduate student, I did not think of these things as much.) A talking bubble has been added over the head of one of the Natives. It reads, "Fuck the Colonizers. Lets take it all back." This meme has 225 positive reactions, split between likes, hearts, and laughter. It was shared 348 times.

In this example of the meme, the text has been slightly changed to "*Three Minutes Later*" rather than *Three Drinks Later*. The interpretation here is that when a group of Natives gets together, it is only a few minutes before they begin talking about rejecting colonizers and reclaiming their stolen lands.

I do not know who made this meme or exactly why they altered the "three drinks later" to read "three minutes later," but I can make an informed guess.

FIGURE 3.5. "Three Minutes Later" meme.

One of the ugliest racist stereotypes of Native Americans are that they are all drunks with a specific, biological weakness for alcohol. The "drunk Indian" is a trope embedded in American culture. I have even been told many times in my life, because it is a commonly told and often repeated myth, that Native Americans are genetically predisposed to alcoholism. I recall as a child hearing in school that the settlers "introduced" alcohol ("firewater") to Indians, who had never consumed it before. Thus, the Indians became hopelessly dependent and would do anything to get their hands on alcohol. This is why, so what I was taught in school goes, the Indigenous peoples made such poor bargains for their land—their addiction and dependency. These are all settler lies. Native Americans produced alcohol before the colonists arrived, and "research" on Native genetics are to be viewed with deep skepticism, particularly in the face of racism that tends to see social problems in whites as individual problems but social problems in BIPOC as racial inferiority. But the issue of alcoholism on Native reservations is complicated because trauma is complicated and its effects on people are complicated. In February 2024, I wrote an opinion piece for *Newsweek* on the devastating crisis of suicide among Native Americans. Some commenters on the article blamed the high suicide rates on Native alcohol addiction. I find the resilience of this myth frustrating and stunning, and I am not alone. So, I can imagine, to avoid this fraught topic and racist stereotyping, the maker altered the meme.

In addition to the alterations to the meme text, there are some interesting feature to this image. The first is that these are not modern Natives, but those from an earlier time. They are portrayed in buckskin clothing and with horses, which suggests that these Natives are from when the Midwest and western United States were not yet colonized. Although this meme does not explicitly date the image, the point here is that Natives have wanted their land back since a time that predates the industrial revolution. This next feature might seem obvious, but it is important to note that throughout the entirety of the existence of the United States, and the earlier colonial era, settlers have always lived on, infiltrated, claimed, and moved onto stolen lands. Manifest Destiny is the ideology in operation for the "settling" of the Midwest and West. The implication here is that Natives have wanted their land back since early in the settler-colonial project—the project that works on the idea that a Judeo-Christian God has destined that the US spread across the American continent(s), bringing democracy and capitalism in its wake. This image harkens to the historic record of the repeated removals and relocations of Native folx since colonization began.

Here again I consider my research questions when I ask what the Indigenous political truth of this meme is, and the answer is that the land is stolen, and the memory of this theft is fresh no matter how long in history we look back. This image is of a community, and the community grows in the image with the arrival of more Natives. The land is threatened here, and the Natives in the image want it back. All of it. There is also implication here in the words "fuck the colonizers" that there was at some point partnership with the colonizers or consideration of the colonizers. And this is a knowing and a truth in Indigenous histories, that the US signed treaty after treaty with Indigenous folx and never honored any of them. The statement "take it all back" suggests that we are finished negotiating and we will take our land from the settlers like they took it from us.

FIGURE 3.6. "Settlers: We Have Culture" meme.

Uhaul Your Ass Back to Europe and Give Us Back the Land.

This is another meme about wanting land back. It has 253 laughter, heart, and "like" reactions and was shared 66 times. This meme repurposes an image on the side of an actual U-Haul moving truck. The U-Haul image is itself an image of a U-Haul truck juxtaposed against an image of settlers in a covered wagon. The caption of this image is "Do-It-Yourself Moving: An American Tradition." The image suggests that people have been moving themselves across distances since colonists settled the West. The page posters shared this image with an added-on banner, "Settlers: We Have Culture," with the Facebook caption, "Uhaul your ass back to Europe and give us back the land."

In other words, settlers suggest that their culture is one of claiming and colonizing North America. The poster's response is that they want the land back and the settlers gone.

When I first saw this image, I thought, "That was really on the side of a U-Haul?" It is the twenty-first century, and colonists still support the history that the Americas were an unoccupied and unsettled land that they could take for themselves. Stealing land is their tradition, apparently, according to this advertisement. I have seen estimates of 2.5 to 20 million people murdered in the Americas, up to 95 percent of the Indigenous populations, by the settler colonization efforts (Diamond 1997). The side of this U-Haul celebrates genocide, and worse, it does so because whoever designed this mural-advertisement was completely ignorant of that history. This image, as much as it infuriates me, is a startlingly accurate capture of what the settlers brought to this land: Manifest Destiny, the nuclear family, patriarchy. This image is known history for settlers, an idealized history, to be sure, as all nostalgic histories are. And that is the rub here—an ad designed to be nostalgic for settlers is the elimination of the rest of us. I do not imagine that is something that U-Haul wants to sell.

Settler-Colonialism and White Feminism

This image macro began as a royalty-free photo on the photo-sharing site Shutterstock in the portfolio of photographer Albina Glisic. In her portfolio, Glisic describes the image as "Smiling happy young woman holding and giving red present to man with white Ribbon close up." I offer my description here: there is a white male on the left side of the image and a white female on the right. The male is holding a gift, wrapped in red, just large enough to need both hands to hold it. The woman is close to him, with one arm wrapped around him, covering his eyes. The photo has been captioned "Settler-Colonialism" over his face, "White Feminism" over her face, and "Continuing the settler project on stolen lands" over the gift she is giving him.

Described simply, this meme, with 155 "like," laughter, and heart reactions and shared 50 times, is an image of a white feminist gifting the continuation of the settler project to settler-colonialism. This meme is a critique of white feminism's concern solely for the issues of white women, and of how their lack of intersectional feminism is a gift to the settler-colonial project. This critique is specifically focused on women, and I argue that the political view and worldview identified here is relevant for at least two reasons: (1) because women are leaders and culture bearers in Indigenous culture and (2) because focusing only on white women's issues does more than just exclude Indigenous women. Through this exclusion, this inability to attend to issues of colonialism

FIGURE 3.7. "Settler-Colonialism" meme.

and capitalism, white women "gift" power to the settler-colonial project. And because it is wrapped in a bow in the image, and therefore care was taken to prettily wrap the package, there is a suggestion here that white women know and do not care about the issues of their Indigenous counterparts.

Before I discuss the next meme, I need to briefly touch on the place of women in Indigenous cultures. While, again, there is not a universal Indian, there are commonalities in American Indigenous experiences. But more significantly, before the settlers arrived, there was no Christian patriarchy, gender binaries, or subjugation of women as property as it is in the settler-colonial paradigm. Settler-colonial ideology works in such a way that whatever the settlers lay their eyes on they can claim. This includes land, water, and women and children. This is not the case for Indigenous cultures. Women were people of leadership in Native culture. Children were valued and raised collectively among the communities and clans. There was no cultivated separation between the genders, whereas in the western world, the genders are so

FIGURE 3.8. "Indigenous Women" meme.

different they might as well be alien to each other. The book *Men Are from Mars, Women Are from Venus* might be a relatively new publication, but its system of beliefs, that men and women are fundamentally different and have rigid gender roles, has deep roots in western cultures. This meme, while critiquing the kind of feminism that only focuses on white, western issues at the expense of all other cultures, is rooted in Native feminism and beliefs and practices surrounding Native women. The meme shown in figure 3.8 is a representation of Native feminism and its place in Native culture.

From my earlier research questions, this meme that has since been deleted from this Facebook group, while about the commonplace and commonly critiqued white feminism, does not work if there is not some understanding of

Indigenous womanhood and Native feminisms. When I originally presented part of this research in the Control Societies Speaker Series at the Annenberg School of the University of Pennsylvania on April 1, 2019, I paired the white feminism meme with this image titled "The front lines of Settler Colonialism," as a representation of Indigenous feminisms. In this image a very large grizzly bear is fighting with a slightly smaller grizzly bear, while a bear cub looks on. The text reads that two-spirit, nonbinary, Indigenous women are "leading the way in the revolution" and "doing other amazing things," while the Indigenous man narrating feels "safe and comforted" by them.

I interrupt my discussion here to note that grizzly bears are sacred relations to Native American peoples across the United States. They are significant in ceremony and culture. The placement of this text in an image of what seems to be a female bear protecting her cub invokes this sacrality as the significance of women in Native American culture as it is superimposed over the female bear and her cub.

I am reminded of the 2019 battle in Congress over the protection of the grizzly bear population. California Republican House member Tom McClintock said then that preserving and protecting the bears is the opposite of western science: "'The science tells us the [bear] population is fully recovered,' McClintock said. 'This bill substitutes emotional, ideological and sentimental biases that are the polar opposite of scientific resource management.'" Natives speaking up for the bears that cannot speak for themselves are the subject of McClintock's words here. All that is valuable to Natives is apparently just emotional, ideological, and sentimental. As if those things are bad. Still, I point out here that men and women are imposed on the bears in this image without equivocation. Here the people and the bears are relations and on equal footing, with the mama bear relating to us the knowledge of Native feminism and defenders of the sacred. Settler colonialism brings us western science and sets it in opposition to the traditional knowings of Native peoples; one is logical, and the other is sentimental. Our government authorizes the hunting of the sacred bears, and there is an eerie echo here between the settler colonialism of the image and the settler colonialism that would continue to allow the unmitigated hunting of animals without consideration of the environmental impact of these acts. If the bears were known as relations to and protectors of the settlers, would barring the hunting of them even be a conversation? I do not think so. But again, for white settlers, it is impossible to step outside of the racialized notions of "others," and those racialized notions always indicate racial inferiority.

I return to the subject of Indigenous feminisms again, because setting these memes together allows us to talk about the contrasts between white feminism and Indigenous feminisms. White feminism supports a white agenda, equality, and autonomy for white women. Indigenous feminists, with their broader notions of gender identity and rejections of the gender binary, allow for an intersectional approach, one that defends us all from settler colonialism.

Elizabeth Warren after She Gets Her DNA Results "Proving" She Is Native

This meme is a side-by-side image macro. The image on the left is known as "imagination SpongeBob" and has SpongeBob spreading his hands, with a rainbow arcing between them. Superimposed over SpongeBob's body is an image of Elizabeth Warren's face smiling brightly with her eyes forward. Superimposed over the rainbow is the word "NATIVE." The right image is a close-up of Warren's face with an unflattering look—her smile is not sincere, and her eyes are looking off to the right instead of at the camera. The background behind her is blurred, suggesting that this image of her face is an extreme close-up. This meme has 298 likes, laughter emojis, and hearts and 195 shares.

The left image is captioned "Elizabeth Warren after she gets her DNA results 'proving' she is native." The right image is captioned "Elizabeth Warren after Indigenous People call out her opportunism and demand she dismantle settler colonialism and capitalism rather than perpetuating colonial eugenics." This response is uniquely Native, in that major media like *CNN*, *Washington Post*, and the *New York Times* focused on the science of DNA testing in their reporting. They wrote extensively about the way tests are interpreted, the folly of constructing one's identity based on the testing, and why the public misunderstands the reading of and the biology behind DNA testing (Kessler 2018; Nelson 2018; Parker 2018; Zimmer 2018).

The argument about whether Elizabeth Warren can claim DNA ancestry over a scientific test is the wrong one, according to this meme. Trump's attacks on Warren when he calls her "Pocahontas" are rhetorically complex, invoking racism with his mockery, sexism, and colonialism, and anti–affirmative action with his focus on her opportunism in identifying as a minority in her past. Identifying as Indigenous is also complex because the very system, blood quantum, used by the federal government does perpetuate colonial eugenics. It decides what percentage "blood" is Indigenous based on the federal

FIGURE 3.9. "Elizabeth Warren" meme.

documentation of your ancestors. For example, if your mother is Indigenous and enrolled in a tribe and your father is Indigenous, but for whatever reason is not on a tribal roll, then blood quantum says you are not "full" Indigenous. Even if you have been raised in an Indigenous family. Even if you personally engage in your own Indigenous community. Even if your phenotype matches what the general notion of Indigenous looks like, dark-skinned, dark-haired, and dark-eyed. This meme's complaint about Warren is not that she is misusing science, on which the mainstream press has focused. Rather, the meme argues that if she were Indigenous, she would be using her political leverage to "dismantle settler colonialism and capitalism."

I thought about this meme deeply, because it does and does not fit my research paradigm. Here, the "what is being protected / what is being threatened" tension is subtle—there is no critique of settlers and their quest to possess land and water. There is nothing about the exploitation of the earth and her resources. If there is the essence of threat and defense here, it refers to Warren herself. It suggests that she needs to take action to protect Indigenous causes—which, by relation, includes land and water—but, unlike the other memes, it does not specify the physical world. This meme does, however, refer obliquely to Indigenous knowings and truths and an Indigenous lifeway. In fact, this meme is specifically about two things: (1) the settler eugenics of DNA and blood quantum (how much "Indian" is Warren?), and (2) how, even if someone is in fact Indigenous, they are charged with assuming the lifeway and epistemologies of Natives. If she believes herself to be Cherokee, Warren *should* demand the return of the sacred land, she *should* defend the sacred water as part of her lifeway and identity.

FIGURE 3.10. "True Spirit of Thanksgiving" meme.

The True Spirit of Thanksgiving

I saved this next meme on the day it was posted. I went back to it in the summer of 2023, and it had since accumulated 319 positive responses and 136 shares. I do not have data on how long it took for this meme to pick up attention, but it is noteworthy that it did. On November 21, 2018, an international news story spread about an American tourist who, even though it is strictly illegal, made his way to the Andaman Islands, deep in the Indian Ocean. The Indian government bans all travel to the remote islands, where several groups of Indigenous peoples live without contact with the rest of the world. When the tourist, who had made more than one attempt to get to the islands, finally stepped foot on shore, he was given warning shouts and gestures. Then the Indigenous peoples of the island shot him with arrows. According to Agence France-Presse (AFP 2018), those who shot the arrows were arrested.

Whoever made this meme screen-captured the AFP story, with its headline and lead photo of the remote islands, and altered it. The original headline was

"US Tourist Killed by Arrow-Shooting Indian Tribe." The maker then crossed out in red the word "tourist" and replaced it with "Christian Missionary." Finally, they added a caption above the headline, which reads, "*The True Spirit of Thanksgiving*." I point again to the date range from the beginning of October through the end of November, which I used for collecting my sample. In 2018, Thanksgiving Day in the United States was Thursday, November 22. Therefore, it was posted to the page the day before Thanksgiving, thus the "Spirit of Thanksgiving" context the meme is situated within.

Although the events from the news story took place in India, the maker of the meme repurposed the word "Indian" in the headline, as the Indigenous peoples of North America are historically referred to as Indians, and in fact many still refer to themselves as Indians, as I have many times referred to myself in this text. Correcting news headlines is a popular image macro meme, often accompanied by the caption "fixed it for you." In this case the fixing is performed by removing "tourist" and replacing it with "Christian Missionary." This meme references the original Thanksgiving as a time when Native Americans were generous with the Pilgrims at a meal to celebrate their first harvest. This meme, then, argues with its satirical humor that instead of sharing a celebratory meal, the Native Americans should have killed the Pilgrims.

I commented on this recently when I was asked by some well-meaning folx for my opinion on Thanksgiving. For the record, I think Thanksgiving is a cultural story that should be told alongside Native American history, including the devastation of colonization on the peoples living in the Americas before they were "settled." It may be a pretty story, but it does not end well for Indigenous peoples, and that reality matters.

An Indigenous Perspective of Political Identity Matters

My story here explores the ways some Indigenous peoples use social media to connect and organize with their shared political views. My story also explores the ways some Indigenous peoples use memes to self-identify and create in-group signs. Finally, my research considers Indigenous knowledge-making practices alongside non-western thought. It identifies and describes Indigenous knowledge-making as experiential, communal, and tied to a relations network. This ability to politically organize Native style is essential given the ongoing efforts to disenfranchise natives and minimize their participation in civic life.

Decolonizing our thinking about political identity and political organizing is, at its heart, rejecting a white, settler-colonial worldview of who people are and how relationships work. Some of the views I have featured here that Indigenous political identity rejects are patriarchy, settler-colonial capitalism, policing raced groups, and racist and genocidal representations of who is and is not Indian and what that looks like to the broader public. Also, here, at stake and inseparable from Indigenous political identity, are land, water, place and space, animals, the environment, and the autonomous bodies of people. I say "inseparable" because, as I repeat often in this book, we cannot simply collapse all North American Indigenous peoples, clans, and tribes (federally recognized or not) into a monolithic group. But there are a few commonalities shared between Indigenous folx, including that humans do not have supremacy over land or animals, that we are defenders rather than consumers of the sacrality of life. There are in this group many memes to choose from. The ones I selected, I think, form a prismatic picture of Indigenous political thought. For example, I chose voter suppression because the problem with requiring specific kinds of identification that might be very difficult to acquire for Native peoples is an ongoing issue that echoes the historic voter suppression of Native peoples. I chose the Pocahontas memes to demonstrate not only racist stereotypes of Indigenous women but also the problems of self-identification and blood quantum alongside political representation (or the lack thereof). I chose the "three minutes later" meme because it reflects this political thought and how it enters our conversations.

I will conclude simply by stating that I chose this story specifically because it is a positive story of Indigenous peoples using social media as one way to support their democratic processes, especially in the face of continued voter suppression of marginalized peoples. It is too simplistic to say that social media will solve voter suppression or that capitalist-oriented technology will consume Indigenous peoples in the continuation of a legacy of colonial consumption. And with respect to stories of Indigenous political interventions using digital technologies, there are perhaps more positive than negative ones. But colonial structures and institutions are difficult to overcome, especially as they are continuing projects. (Like decolonizing ourselves is also a continuing project.) And social networks and networked technologies are integrated closely into our personal and civic lives. Telling Indigenous stories of and through these technologies is a project of writing the narrative—of writing over spaces where we were erased. I hope these stories have given you some awareness of the issues at hand and the stakes for Indigenous futures.

4

Jeffrey Veregge

A Story of Relations

Why Rabbit Is So Lean

Bear and Rabbit met at Bear's cave and had conversation just like our folks.

"How your folks getting along, Bear?" asked Rabbit.

"Fine. How your folks?"

"All well," was the reply. Bear and Rabbit, just like folks, talked all morning about nothing much.

Finally Rabbit said, "I must go home."

"Wait till I have dinner," Bear urged. He cut a piece off his side and fried it for dinner.

"Come see me, Bear," said Rabbit as he started to leave.

"Where do you live, Rabbit?"

"On old field with tall corn."

One day when Bear was walking along, he remembered Rabbit and invited him to come see him.

Bear walked through tall corn but could not find Rabbit until he stepped on him and Rabbit went, "Squash, squash!"

Bead said, "Hok, hok, hok! How you getting along, Rabbit?"

"All right. How you, Bear?"

"Fine!"

> "I'll get dinner," said Rabbit. He got a big knife and cut a piece off his side just like Bear did.
> He fried it and gave it to Bear.
> Rabbit's side is still lean. (Edwards n.d.)

This is the story of how I met and made meaning together with Indigenous artist Jeffrey Veregge in September 2020.[1] But first, before I get into that 2020 conversation, I need to go back a few years. On April 13, 2013, the comics, science fiction, and fantasy website, io9, posted a story by Annalee Newitz about a S'Klallam artist, Jeffrey Veregge, who drew superheroes in traditional Indigenous Pacific Northwestern form line art. I enjoy science fiction and Indigenous art, and the *io9* story about an artist who combined the two made a lasting impression on me. Jeffrey's[2] work is beautiful, and his application of his Indigenous style to popular cultural icons opened up for me new possibilities for seeing and reading those icons through Native eyes. My admiration is not just because I could imagine Indigenous peoples as superheroes, but also that superhero stories can also be our stories; they are not limited to white audiences. My favorite image of Jeffrey's that was shared with *io9* is of a black, white, and gray form line drawing of Batman midleap with his wings extended behind him, but they are all wonderful. As a result, I began casually following his career by connecting with him on Twitter and Facebook.

Jeffrey's work has been seen in comics and as murals that are part of the Smithsonian's permanent collections. He designs public murals. He shows in galleries. He has provided cover art for more than one hundred comic books, and he has collaborated with many other graphic novelists and artists, including the Indigenous fiction writer Rebecca Roanhorse. As I write this chapter in October 2020, Veregge is partnering with Marvel to release his art on a poster and in cover variants in November to honor Native American Heritage Month in 2020 (Marvel 2020).

I asked Jeffrey Veregge to do a videoconference interview with me to discuss his Indigenous identity, his art, his career, and his activism. Initially, my goal here, because I am writing about digital rhetoric, was to get a sense of how he constructs his digital identity given the steep rise of his fan base since that first *io9* article in 2013. I had also noticed in recent years that he had answered questions during interviews from journalists about Indigenous issues and his faith, and I wanted to know more about how he had become a kind of activist through his presence on digital platforms. If I am honest, I wanted to talk to an Indigenous person who is composing science fiction and fantasy, composing the future, while keeping his feet grounded in traditional methods, which I think is a fantastic representation of Indigenous peoples today. Like I said earlier, my goal *was* to look at Veregge's digital rhetoric and his popularity among digital audiences. But as I was writing this chapter, it occurred to me that the structure of my interview and my approach to relaying it would have a strong impact on how the reader interpreted both my intent and my findings, which was not my goal. I wanted my readers to meet Jeffrey. I realized then that my goal was to make space for Jeffrey and to do the meaning-making together in a way that makes Jeffrey himself and his work the focus of this chapter, rather than focusing on commonly accepted scholarly organizational structure and interpretation used in reporting an interview. With this book, I am not trying to emulate western structure. My intention for this book is always to rupture rather than recreate.

As I mentioned in writing about Indigenous theory in chapter 1, interviewing as a research method is fraught due to the primacy of western methods. For example, while drafting, I initially framed this chapter out with a research question, and then I began to write it with interpretations and observations about Jeffrey's answers to each question. Then I stopped and realized I had already been pulled into doing western scholarship with its underlying assumptions about how research should be organized, conducted, and reported. For example, I reference here *Contemporary Feminist Research from Theory to Practice* by Patricia Leavy and Anne Harris (2019), who design and describe a cohesive research interview plan that urges readers to think about "whose perspectives, experiences, and voices have been left out" when choosing sampling strategies and interview strategies for those participants (140–42). While this feminist approach strives to make visible marginalized voices, as methodology, it is less appropriate for an Indigenous interview, because the power here still lies with the researcher in the structuring and question-formulation, and not with the storyteller or participant (Kovach 2009, 125). I wanted to share my meeting with Jeffrey and reflect him in a good way. When writing the interview questions many months before, I designed the interview as a conversation between Jeffrey and me in an Indigenous framework. I wrote very unstructured questions that would allow us to engage in the back-and-forth appropriate to storytelling methods. Leaning on western methods and research organization was something I did not plan to do; I only slipped into them during the writing process after the interview was over. I describe this here as a cautionary tale, because if we are to engage in decolonizing practices, we need to think carefully about the hierarchy of our methods and our questions, and all the question guidelines that go with structured interviewing. It is very easy to slip from all that hierarchy into an unbalanced power relationship that serves the goals of the interviewer rather than what serves the partner.

Jeffrey and I talked on September 15, 2020, over Zoom from our separate homes. Although this book is not about Zoom, I note here that the context of the pandemic never leaves this text, and that while all of my stories are situated in the place, space, and time of the pandemic, this interview and story were also set in the shadow of the 2020 presidential election. It is one thing to connect on Zoom because you prefer it as a mode of conversation. It is something else to be restricted to Zoom in your ability to meet and talk together, because you cannot travel. It feels restrictive, and that the restriction colors our context should be considered here. It is more difficult over Zoom to read facial expressions and body language. The interviewer and participant are confined to the frame of the screen as if life is not always moving all around us. We can talk, but we cannot shake hands or pass back and forth objects or examples. We cannot sit with our heads together poring over images and sharing process. Online conferencing is a powerful tool, but it has its limitations. Perhaps someday our paths will cross in person. Finally, I sent Jeffrey the questions ahead of time, and they are brief and unstructured, so I will include them here instead of listing them in an appendix:

- Could you share a little about yourself?
- Could you describe yourself and your artistry?
- You're a public voice for Indigenous peoples and their issues. Can you talk about that?
- You're on Facebook, Twitter, Instagram, and YouTube. How do these digital platforms support your work?
- Is there anything about these platforms that undermine or impact your work negatively?
- Do you have any advice for other Indigenous activists and artists working in this digital world?

My first question asked Jeffrey to speak about himself, to introduce himself. The following quotes are from our interview, which I am weaving into a story. As mentioned, the object here is for you to meet Jeffrey as an Indigenous artist working with his hands and online. I hope this weaving gives readers a sense of how Veregge works through his many identities, in Indian Country, in geek and nerd culture, in the greater artistic communities in which he moves and grows, and in his spiritual life. The interview begins with hellos and a refresher from me about his consent and how he can withdraw it at any time. I explain that he can review my draft of this chapter and contribute changes and corrections. He assures me that he is an old hand at these interviews now and that he has pretty much found his flow with them. I also pause here to comment that the interview was friendly, with shared laughter and shared experience. Jeffrey, with his warm smile, curious eyes, and expressive hands, is someone I immediately like and someone I want to know better. I ask him if I may begin recording and he agrees. He opens then with a question:

VEREGGE: Okay, okay, how do you pronounce your last name?
TEKOBBE: Tek-Oh-beh.
VEREGGE: Tekobbe?
TEKOBBE: Yes. It's Chahta Nation. From my mom.
VEREGGE: Yeah, yeah, it's totally cool. Right on.

Here in the interview, I smile happily, because this ritual is familiar to me: we begin with the identification of our indigeneity and our families. I then ask my first question, "Could your share a little about yourself?"

VEREGGE: My name is Jeffrey Veregge. I'm from the Port Gamble S'Klallam tribe. I'm of not only S'Klallam here, I'm an enrolled S'Klallam tribal member, but I'm also of Squamish and Duwamish heritage as well as Irish and German on my dad's side. My dad is the non-Native, my mom's Native. I grew up on our reservation out there—the Port Gamble S'Klallam tribe out there known as Little Boston in Washington. Not a great big place, but it was a fun place to grow up.

There are two things about this moment that I want to talk about. The first is the need for and the ritual of building connection. Jeffrey and I do the work of building this connection as we exchange names, places, and people. This exchange is traditional between Native folks, we share names and family names to verify relationship and situate ourselves in community (Kovach 2009; Wilson 2008). Here, during our conversation, I not only answer his question with the pronunciation of my name, I also explain that it is a Chahta name I inherit through my mother, and that I am enrolled in the Chahta Nation of Oklahoma, who were, before the removal, Chahta living in western Alabama and Mississippi. I shared with Jeffrey that my institution, the University of Alabama, is also in western Alabama, and that my connection to my job is interwoven with my connection to the land. Jeffrey nods and tells me this is cool, and I smile again because I think it is cool too. It is something personal and powerful to me that I seldom share, but I feel a kinship with someone who also identifies with place and land.

Jeffrey then introduces himself and where he and his people are from. Through this exchange, we describe our origins and heritage—a connection through identification. We are both Natives of mixed families and Native spaces, connected to peoples and land through this knowing of who we are and where we come from. We approach each other in good relations, naming and situating ourselves in traditional ways. I explain I am somewhat anxious as I work as an Indigenous woman, because doing the work means opening myself up to critique and the cultural need to place people in racial boxes, as in, because I am an Indigenous person writing in public, then, as in my earlier discussion of Powell's "prime" narrative, I am representative of all Indigenous people's views and beliefs. And Indigenous folx are already agitated by and sensitive to a history where they are talked over and treated condescendingly as less intellectually sophisticated. Worse than being put in a box neither Jeffrey nor I want is the concern that something we say will cause offense in Indian Country. We always respect and value the voices of our elders. I ask that the reader keep in mind the risks of doing indigeneity here in public. I will talk more about this later in this chapter, but to begin with, please accept that there is always in the background the imperative to represent ourselves and our people in respectful ways, and that missteps can have consequences that ripple beyond the moment.

The second thing I need to talk about is the ritual of introduction in relation to relationships and connections. Linda Tuhiwai Smith writes, "Building networks is about building knowledge and data bases which are based on the principles of relationships and connections. . . . People's names are passed on and introductions are used to bring new members into the network. The face-to-face encounter is about checking out an individual's credentials, not just their political credentials but their personalities and spirit. Networking by indigenous peoples is a form of resistance . . . establishing trust is an important feature" (2012, 157–58). Although it was over Zoom, Jeffrey and I made a face-to-face connection. We exchanged credentials, we assessed personalities, we laughed together, and we established a connection to complete the interview. My intention here, and because he asked, is to send Jeffrey a copy of this book once it is published. I hope that our connection will grow over time, because I think his work and his voice are important.

Continuing his introduction, Jeffrey speaks about his background and education. He tells me first about his western education before his Indigenous apprenticeship, which I think is probably unavoidable given my positionality as an academic. I come to Jeffrey representing myself and my project but also with the ethos of my university background. In some ways, I represent western education and its monolithic colonialism in this interview. It is a positionality that I cannot really remove from this space of ours, even as I do not want to approach Jeffrey as a subject–object. I also want to point out here that he lists his education chronologically, and he was educated in western schools before he moved on to studying art:

> VEREGGE: I went to school, graduated high school in nineteen ninety-two, went to art school, graduated there in two thousand, so I took a little time for that after I graduated with an honors degree in industrial design from Art Institute of Seattle. And then after, that's what got me into form line design.

His discussion about form line design allows me to bring up an issue that many Native Americans have, and that is having a cultural connection to traditional ways of meaning-making, but without access to detailed or deep knowledge of those ways. Jeffrey here mentions that he "faked a lot of stuff," and I do not want to directly disagree with him or his experiences, but I will discuss how he worded this. Jeffrey is not "fake" S'Klallam or "fake" anything. In his own account, his own story, he explains that he came up on the Port Gamble reservation in the Pacific Northwest. He is very real. "Fake" here, then, speaks to a tension that has troubled generations of Indigenous folx since colonialization. With fewer resources to put behind a cultural education system, it can be difficult to access those traditional knowledges and arts. And western education is normative for nearly all people in the US. I argue here that Indigenous art forms do not have the benefit of teaching and learning in the American public school system like other forms of art, other histories, and other literatures. What Jeffrey does then is imitate shapes and ideas that he has seen, which as a literacy practice is one way we can learn, by imitation. I do acknowledge that he had not been apprenticed early in his career and that the retention of culture is critically important. Form line lives in Jeffrey:

> VEREGGE: Really, what I do now—I did a logo for my tribe, and it was a form line design. It was for their family services, and they liked it and I thought it looked good. But I also knew that I faked a lot of the stuff. I just used shapes, and seen repeated elements. And being that I just graduated art school, I knew that there's other artists out there that would look at my work and say, oh, he's full of shit, that he doesn't know

what he's doing. So, I went to my cousin, who was a cultural liaison of the tribe and was friends with Tsimshian master carver artist David Boxley. And I asked her if she could make an introduction for me, and she did. David asked me why I want to study with him. I said he's the best that I knew of, and that he had work in Disney World. He travels all over the place. He did beautiful work for my own tribe. And I just thought that he'd be the best person for me to learn from. So, the designs that I do are more Alaskan-based shapes than you would find traditionally here in the Seattle area, although the designs, my own designs, have taken from Salish work, as well as Alaskan Tsimshian style as well. But it's not a pure form of either; it's a hybrid of both, along with the traditional art and design elements that I've learned over the years.

Here Jeffrey is talking with his hands, and I am nodding my head during this part of the conversation, because I share this feeling of faking it until I was able to learn it. So, I too needed to reach out to Native mentors to learn to think and write like a Native person. For me, I had the elders and members of the American Indian Caucus at the Conference on College Composition and Communication (CCCC) to help me explore Indigenous writing and research methods, decolonization, and indigeneity. When Jeffrey reaches out for a mentor, he asks a cousin to make introductions for him; these introductions are connections between networks of relations, where Jeffrey's people intersect with Boxley's people. And by *people*, I mean the thick meaning of "people," as in our families and relations in our communities, where we have constructed thick bonds. I am slowly and carefully learning my own culture. The traditional greeting that I use when I give conference and invited talks was translated for me by tribal elders. When I am able, I take Choctaw language classes. I read books about and written by Choctaw women and storytellers. I follow the Choctaw Nation YouTube channel for videos of governance meetings and cultural programming. I keep up with the Nation's website. These digital platforms and content are helping to build Choctaw community outside of the limits of the reservation. And many tribal communities are doing the same for their people with their own traditions and practices. These platforms give us access to culture we would otherwise not learn.

Across those different communities are woven the different cultures and identifications of peoples. There is not one Indigenous mode of art, or Indigenous mode of research. There are as many modes as there are cultures, and then artists and thinkers within cultures reflect their own uniqueness—not unlike western artists. There is no homogeneity across the country, and while non-Native audiences may recognize a piece of clothing, jewelry, pottery, story, or sculpture as a Native style, it takes knowledge and familiarity to identify those specific styles to cultures. And as I mentioned above, there is pressure on Indigenous public voices to perform Indianness for the greater culture. There can be pressure to speak to the experience of "all" Indians. Here, Jeffrey is delineating his relations networks and how he connects to his mentor, but he's also saying that he inherits his style from multiple communities, and referencing them, educating me and the readers here with knowings we would not otherwise have access to:

VEREGGE: After I worked with David, I worked at a marketing agency for eleven years. I started as an intern, working my way up to the lead designer studio manager. And did, worked everything from mailings that people call junk mail to websites, catalogs, exhibit booths, mostly for Christian nonprofits. Was a great learning experience for me. I learned so much about the print process and the demands of working on multiple projects with tight deadlines all the time, working and communicating with clients. It really prepared me for what I'm doing now. After I left Masterworks—I was laid off. We had suffered some financial troubles, and I was laid off, and I was trying to figure out what I was going to do. And I just started doing the native art stuff and comic books. And I just got my first comic book gig with IDW [Publishing] and another comic artist who I greatly admire, Tim Sale. I had been talking to him and he told me to go for it. He says if you got severance pay and you have, you already got your foot in the door, you go for it, he says, it can be hard, but go ahead and try to do it. And so, my other friend said the same thing and it took some time and work. It meant I had to do a lot of late nights, take a lot of other jobs as well as comic book work, and I still do that.

But as here we are, over a hundred comic books later and all the other stuff that people know—the Smithsonian exhibit, the various murals are projects that I've got to lend my skills to, it's been amazing. I've got to meet so many of my heroes and work with not only them but their families, and, for a nerd, especially for a shy geek growing up on the reservation, this was, I wish I could sometimes go back and tell him.

We laughed here together a bit ruefully. How common a sentiment is it to wish we could share some wisdom with our younger selves? I have heard this often, and in fact, I think I have said it often myself. But I think it is particularly commonplace in geek and nerd culture, because tampering with a timeline is very popular science fiction and fantasy trope. Returning to a younger version of yourself seems more possible when there are movies, books, and comics that play with time travel and alternate universes to tell stories. I will add here that in Indigenous cultures there are creatures and people who move between bodies and worlds, who show up and give advice or lend support or bring messages and then disappear again. I have written several examples in the earlier chapters of this book. I include this to draw the reader's attention to the wide range of popular stories that are told about and around the fluidity of time, like my telling of the Corn Goddess or 1985's *Back to the Future*. To wish you could return to your younger, less secure, less socially accepted self and tell them that they are going to be fine is something Jeffrey and I have in common, and which our cultures we identify with have in common.

> VEREGGE: But with all I get to do and see what I get to see. It's been great, it hasn't been easy. Like I say, it's a lot of hard work and a lot of. . . . And humility, you've got to make sure that you have a thick skin, and you've got to be humble about all the stuff, you've got to take feedback positive and negative, both equally. And just not your failure and just always see: no is a no to that project or that pitch, not necessarily to your work, as, I've heard quite a bit.
>
> TEKOBBE: I can, I can relate to that. As an academic, you have to take the—you have to put your work out there, which is at first really scary. Right?
>
> VEREGGE: Yeah. Yeah.
>
> TEKOBBE: Because you put your work out there and then people have opinions [laughter together]. Yeah, I get that.

Here, Jeffrey and I are connecting across a common need to accept and process criticism as an artist and an academic writer. Criticism is both the same and different for public voices of Native American communities. First, there is always the question of whether to take this work out of the context of the community. Next, it is likely a relations network has opinions about that, which as a member you are motivated to at least listen to, if not entirely heed. The capitalist drive to demonstrate personal greatness applies less significance than a Native need to do what is best for relations and community. When I write, I find myself asking, what are the concerns of my family and communities? Then I ask, what are the concerns of my fellow writers? Then I ask, how does this writing work for the public who does not know either the Native context or the artistic context? The question "do I like this?" is weighed against whatever aesthetic the reader appreciates, and does not consider the complex context within which the work was produced. I was recently told in a Facebook group that I cannot have an opinion on Choctaw issues, because I was not living on the reservation as part of that community. This, despite all the efforts the Nation makes with its online presence to unite us across the boundaries of geography, is a fact. I don't live there. And the rebuke stung.

While I am an admirer of Jeffrey's entire body of work, I am particularly fond of one piece, "Seduction of the Moon," at Stonington Gallery in Seattle, Washington, which displays works from the Pacific Northwest and Alaska. I am excited here to ask about this piece that I loved the first time I saw it. "Seduction of the Moon" is a rectangular piece with the vivid blue moon above the deep blue-green waves of the sea, and then the Moon again as a reflection of his face on the water. Moon has eyes and red lips, and Octopus Woman is red and swimming with one of her tentacles brushing the lips of the reflected Moon. It is dramatic and lovely. At this point in our conversation, I tell Jeffrey that I want this piece, and I asked him the origins of the work. Jeffrey offers this explanation:

TEKOBBE: You have a piece that I just saw about the Seduction of the Moon. Is that right?

VEREGGE: Yeah.

TEKOBBE: Yeah, yeah. That one's so beautiful. I love it so much.

VEREGGE: Yeah. Thank you. And that was my first. I tend to stay away from traditional-type storytelling, but the more public works that I get, the more I find that I want to tell my own Native stories and make some S'Klallam tales, so the story of Octopus Woman is an ongoing project. I'm creating a twenty-four-foot sculptural piece in Ballard, Washington, that will be on their waterfront. That will be an Octopus Woman piece.

TEKOBBE: Congratulations!

VEREGGE: And, but, the story, that's only part of the story, the, the rest of the story is that I was trying to think that most myths, you know, especially Native tales, they always have an explanation for something. I thought, what would be something cool to explain? And I thought phosphorous. I don't know if you've seen that in the water and how it glows from the—during the waves in the saltwater, so I thought, what if the Moon fell in love with her and he started blowing her kisses? And as you blow your kisses, he blew her so many of them that the water started absorbing, the creatures in the water, and they now glow as a result of that.

TEKOBBE: I love that.

VEREGGE: Thanks.

TEKOBBE: I'm excited for your sculpture, too. I didn't know you were. So, are you working in sculpture? How's that going?

VEREGGE: It's going OK. I'm still in the—we're still in the design process with that. They—I'm getting closer to finally getting the final design ready for that. So, I'm really looking forward to it. It's my first, as my first really big sculptural piece that I've done. I've designed a waterfall or, not a waterfall, a water piece monument for the Muckleshoot Tribe, for their Veteran's Park. But this is the first time that I'm really designing something that's almost a totem, that would be kind of a modern totem pole, I guess you would say.

I asked my next question, "You're a public voice for Indigenous peoples and their issues. Can you talk about that?" Even as Jeffrey is in the process of designing public sculptures and public projects, even as he does interviews where he speaks to Indigenous issues, Jeffrey tells me he did not intend at the start to be an advocate for his people or Native Americans. And I believe him. I think, even as most Native Americans are not taught our histories or cultures, a number of our issues still make national news when they cross over into white concerns, like the Dakota Access Pipline (DAPL) or whether or not Elizabeth Warren is a "fake" Cherokee. With so few Natives in the United States, when you have one in front of you, it is probably pretty compelling to want to try to get your questions answered. Jeffrey is asked a lot of questions:

VEREGGE: You know, it's interesting. I didn't set out for any of that, actually, when I was doing artwork previously. I did create a lot of work the first couple of years, when my first earlier gallery works were all very thought-motivated and reaction-motivated, and wanted to get people thinking about things that were happening in Indian country. And even though I was selling the art and was successful, I was getting commissions, it just, it wasn't really me. And so, that's why I started doing the comic book art, because that's that's who I am.

And, a result of that, the result of people taking notice of the work. I've been asked more and more for my thoughts on other issues, which is, kind of—my work and the notoriety of it has allowed me a platform to share some things that have been on my mind over the last forty-six years. So, it's been an unexpected but a very cool thing to happen.

As it is, I remind myself, I try to remind myself who I am and who I represent, because I represent my tribe. I always carry them with me. Everything that I do is a result of growing up down there. And so, I try

to carry that. When I speak about things, I try to speak to things that I
know of. If there's a cause or there is a something going on that I'm not
quite sure—I have put my foot in my mouth in the past. So, I've learned
lessons there not to talk about things I'm not really aware of.

Here we both laughed full-body laughs, because when he finishes this comment, I acknowledge that putting my foot in my mouth with this book project is perhaps my greatest fear, a fear that has made my progress slow, and for the reasons Jeffrey elucidates here. Everything I do represents my relations—my family and my networks. As Jeffrey says, we carry them with us. When you speak, it is within a complex context of those relations. You are a product of them, and as tradition would have it, you carry them with you, with those from whom you are made. White audiences often imagine that they have a unique identity, a rugged individualism of the white American ideal (Ratcliffe 2005, 111–13). Where white is defined as the norm, and nonwhite is the deviation, white has rugged individualism, while nonwhite is reduced to generic and faceless groups. Therefore, when a nonwhite person, a Native, is speaking in a public way, they do not have access to individualism. Rather, they are the reductive voice of the group. So here, there is a dual bind: we cannot risk misrepresenting our relations, but if we do manage to speak correctly and appropriately for the occasion, we will be stripped of our context anyway as we fall into the generic wash of nonwhiteness, where we must represent ALL of the peoples of Native America, not just ourselves and our (small) networks.

Still, the rise of interest in Jeffrey's work and his career has brought about opportunities for interviews and for public platforms, as he describes here, and those opportunities are important, because his might be the only Native voice his audience comes into contact with:

VEREGGE: But here, you know, I've had a chance to talk about stereotypes.
Native stereotypes are, have been, well, around since the first settlers.
We've always been the villains and, you know, we're just finally now starting to turn that curve. We are not the villains, we're not these parodies
of. . . . These dime store shows that on the shelves that you see in there,
and we're not these western bad-wig-wearing Indians that are whooping and hollering and trying to kill John Wayne. We're much more, or
much, much more than this, and that's allowed me a chance to comment
on those types of situations, it's allowed me to talk about things like logo
design for the Cleveland Indians and the Washington Redskins, that
which from an artist standpoint, as Native standpoint—like, I hate Chief
Wahoo.[3] I was so glad to see this. I'm a huge baseball fan, and just see

Cleveland do that. That was a huge step for Native America, as well as, you know, that, seeing the Redskins doing what they're doing. It's been awesome. So being able to talk on that, and then the other side of it is that I could share a lot of things that are important to me, things like the environment and climate change and science. I like that. My hope is that if kids tune in, or that they not only see my art or my website or my social media, as they see things that make them go, oh, yeah, I like that, different types of science, earth sciences, astronomy, NASA, all that sort of thing has allowed me to share my perspectives on that and show that, a much more rounded individual. If that answers your question, I imagine I probably I do have a tendency to ramble on, so . . .

But Jeffrey does not ramble. And I tell him this, that to my ears this is not a ramble but a story of resistance. Packed into this statement is passion for Native identities and Native issues. Jeffrey speaks about settler-colonial representation of Native lives and settler-colonial adoption of racist iconography of Indigenous peoples to represent sports teams. These caricatures are offensive, because they flatten the complex identities, the thick knowing of Native Americans into flat jokes with a single punch line. He then moves on to how the opportunities to talk about these issues in public gives him a platform to move forward into other issues of Native concern, like environmentalism and climate change. In line with his passion for science and geek and nerd culture, Jeffrey throughout this Zoom interview is wearing a baseball cap and a T-shirt with iconography from the *Star Trek* television series, *Star Trek: Discovery* (2017–present). I compliment his shirt and tell him how much I like futurism in fiction. I transition to the comment he made about environmentalism, and I tell him how with the recent fires in the West,[4] I am very concerned about the environment, climate change, and the denialism that seems to come into play whenever we try to publicly discuss these issues. I brought up the fires specifically because I have recently been reading about trends in land management that seek the help of Indigenous peoples' histories and practices to manage forests without catastrophic fires (NPS 2023):

VEREGGE: It, you know, all this stuff, 2020, has been this year that it really makes you think, like, it's brought all our sins to the forefront. Our sins of racial inequality injustice, we're seeing that—we're seeing a piece of that—and we're getting changed, that the environment itself, the fires, smoke, all that as a result of our consumerism and our need for quick and easy solutions. And then we have COVID, which is another pride, where people aren't wearing a mask and they're spreading it, so I think that, that's one thing 2020 is, really. It's given us a glimpse of: we can go down this path, or we can go down that path, and my hope and it seems to me that everybody else's hope is that we go, we right the ship. We, we're making progress, and hopefully everybody else will catch up.

We paused here for a moment with the intensity of these words. I have made this argument and am making this argument at several other points in this book, that late-stage capitalism, that peak consumerism, is responsible for the current trends in the destruction of Indigenous lives, communities, and ways of knowing. While not all Indigenous peoples are monolithically environmentalists, most Indigenous peoples have a traditional relationship to the Earth, the Sky, and the Water. The Water Is Life movement is a literal statement that water is alive, and water is the life of the people. The negative language of pride and sin is not uncommon in Native American spirituality. Some of it, of course, is influenced by Christianity, and some of it is older and grounded in traditional stories and lessons, including some of those I have shared in this book. But here, Jeffrey signals his spiritual practices, as he identified COVID deniers as individualistic and selfish, and therefore people whose pride harms the community. He speaks of racial injustice rooted in the practices of empire that are literally damaging the planet. And once again, Jeffrey and I are talking about our context and our exigence, that this very moment as we speak, fires spread out of control, racial injustice is at the forefront of American discourse, North America has not yet learned our traditional lessons, and we are all paying for the consequences.

Jeffrey continues talking about how sharing his work brings awareness to Native issues. With my question "You're on Facebook, Twitter, Instagram, and YouTube. How do these digital platforms support your work?" I ask about his career and how it seems to grow along the lines of the increasing influence of social media platforms. He agrees and argues that social media interactions provide real-time audience feedback for an artist, and then he loops back to networks. I write a lot about networks. I write about network theory, and I write about relations networks. These two frameworks have some similarities, and in this discussion, they blur a bit, as Jeffrey is talking about digital networks, but also the meeting of persons and the making of relationships that is Indigenous knowledge and kin-making.

VEREGGE: I get people aware. I share my work. I never show how I make the art. I don't like showing that. Sometimes I think that it ruins the illusion, that it's cooler to see the finished piece. But I use it just to communicate, and to me, that, today as an artist, is, should be, the easiest for anyone, that you have these platforms that are not just local, but it's global and it's instantaneous. And you can see reactions. You can gauge interest that way in ways that could never have been done. The work that I've done, everything that I've accomplished has been a result of being online and sharing that, sharing the right places, making the right contacts, talking to the right people, sharing the right works, and taking advantage of every situation that comes my way with that, just using a lot of the practices that I learned in the marketing agency I had, about how to represent myself and how to engage with people online. It has been, it's been priceless, I guess, to say, for me, this all started as a result of *io9* picking up my website and sharing my art eight years ago. Yeah, and as a result of that, it just bloomed from there.

As I wrote in the beginning of this chapter, on April 2013, the website *io9*, which describes itself as having "all the top news about comics, Sci Fi, and fantasy," featured Jeffrey's artwork in a piece titled "Epic Superhero Art in a Traditional Native American Style" (Newitz). While Jeffrey had been a working artist for more than ten years at the time, the website brought international audiences to his work. While he'd been on Facebook and Twitter for a couple of years at this point, his circle of influence on social media grew much larger almost overnight. I ask Jeffrey to describe how the rise of social media and his rise in popularity happened almost in a collaborative arc:

VEREGGE: Yeah. And that, and that's been great, that the social media has allowed that. It allows me to stay relevant, you know, and to keep pushing myself. It not only allows me to share my work, allows me to see other people's work, and to remind myself that I'm only one of thousands, millions of artists, and I'm lucky enough to do what I do. But in order to keep doing what I need to do, I got to keep pushing myself. So, I use that. That's another way. Just kind of motivation.

It has been really cool to see people that became fans when I did my *io9* show when *io9* shared my work. And I got a lot of followers, and some have become dear friends as a result of that. But they've seen the progress just from that work to where I'm at now. And they always are telling me how you can't believe how fast it's gone. But a lot of them don't realize that before that there was all the work that I had to do to get to that point. You know, it's never A to B. A to B can take as long as it needs to get. For me, I had to paint. I had to do art. I had to study art for a couple of years. And not just, not just art school. I'm talking about on my own. I had to go back to art history, who do I like? Why do I like them? And what is it about their work that strikes me, and what is the ideas behind that, that I like? So, it's involving all that sort of stuff that has really led to this moment that people tend to see the immediate stuff, the comic book work, but not realizing that, yeah, that I graduated art school to get to here. So, twenty years.

I ask Jeffrey about any downsides to his social media presence, with "Is there anything about these platforms that undermine or impact your work negatively?" I ask this question to complete the circle on a digital life, thinking about what is good, what is bad, and what is complicated. I want to know about how his digital life, especially in its difference for Natives, who are seen as representative of all Natives, impacts his whole life and his identity as an artist, father, husband, man of faith, storyteller, and member of the S'Klallam.

VEREGGE: Yeah, you can definitely take away your time from working. It's a distraction, definitely, definitely lowers your, my, already short attention span, because now you find yourself checking things. My wife, since COVID started, she had me offline and more present at home, much more present. It was getting to be a real problem for me, just being honest, going back and forth between Facebook and Twitter and constantly being involved, feeling like I had to post everything and comment on everything. And since last March, I've been, really drastically cut that back, and it's been great to be home, be able to do things, especially now with my workload. It makes it a lot easier for me to not do that there.

It's hard, though, I mean it's like anything else you get on a, like, on it, and it's like, it's a decent reward, so you start getting used to seeing people like things and you want that, you like that feeling and you can get addicted to that really quickly. So, what I try telling everybody else is, if they're getting into the artwork or, social media is a great tool, but don't make that just about that, because I can get to that. I could have easily gotten to that point where I just wanted to be online all the time, and I wasn't focusing on what I should be doing, which is my work. And that's a first, foremost thing. If that's the reason why I am, where I'm at, work that I create. And that's what my wife reminded me. That's what you need to be doing. You got to remind everybody, why not necessarily share all your thoughts? You say that's not a bad thing, but just remember that it's artwork that got you here, that you need to.

We both expressed frustration with all of the channels of communication that come through us at any given time of the day. I pointed out that I am always online, and I generally work this way because it is how I connect with colleagues across distances. While I am working and connecting to collaborators, across my feeds come frustrating posts, and I have to tell myself that it is okay not to respond. It is okay to remind myself that I cannot respond to everyone. I think this is where my concept of relations networks and my concept of digital networks really come into conflict, in that (1) I have a sense of responsibility to my relations (and in my relations group are family, friends, and colleagues) to communicate, because through those relationships is how I think, and (2) I am part of a greater persistent digital network, where information flows through me but does not necessarily require my response or come with the same commitment to communication that my relations networks have. Decolonizing myself would focus my commitments to my relations and learning to say no to the other things that compete for my attention, but that is an ongoing project. Here, Jeffrey describes his response to these persistent messages and communications:

> VEREGGE: Oh, yeah, totally that it's, people don't get it, and I feel bad for friends that I talk to and colleagues likewise. They start messaging you, and you start talking, and you don't realize this takes up a lot of time, they don't realize the amount of time to that conversation, that I finished the job or did something you know and—So, it's really hard, and that's one that I have to always remind myself, because I can get drawn into some really bad, long, complex thought-out conversations.
>
> Facebook especially was a really good thing to get off of. I'm a very... not a big fan of our president,[5] so you can imagine that, getting some pretty good discussions there on this, because I really try to let people have their own opinions or whatever.

We agreed here that in respecting others' opinions, we draw a line at the racial and social injustices we talked about earlier in the interview. Jeffrey and I both agree that racism is a clear and distinct issue for us, without any gray areas. We cannot see people as simply, flatly racist, but we see the way that racism is embedded and intertwined in so many ways of knowing and thinking that it cannot be extracted from the representation of the person. It is one thing to say we accept different viewpoints than our own and another thing to entertain ideas and words that reject our values. I tell him this, that I cannot do Trumpism, because I cannot do racism and anti-environmentalism, and this has cost me some relationships.

> VEREGGE: Yeah. And I've lost, I've lost followers as a result of talking about my political views and even my spiritual views. OK, so I believe in God.

I believe in Jesus. I am proud of that. I pray, read my Bible every day. However, I do swear, I do smoke pot from time to time, and I got a pretty twisted sense of humor. So, and I believe in equal rights. I believe in everybody should be loved, everybody should have, this is, to me, LGBT. Nothing wrong with that. I have very many close friends of family and I, I get—people often associate thinking that because I'm somebody who has spirituality, that I am right, that I could be considered right-wing. Like, now, it's not even close to that.

We have a discussion of spirituality between us, both of us identifying with the social justice values of liberal Christianity, and how we do not budge on those values, even values we think we share with other people, but we instead discover things about those in our networks that we have to step back from. Spirituality is for me a vexing subject, because it is not something that often comes up in white academic scholarship in my fields of study. Spirituality and talking about it, in my experience, has made other scholars in the conversation uncomfortable, because western rationality rejects the spiritual. African American rhetorics has a substantial spiritual component, particularly as it is related to rhetorics of resistance and justice. Native American rhetorics also incorporates the spiritual in complex and significant ways. Also, given the Christianization efforts of the settler-colonial project, it is commonplace to find Christianity practiced among Natives, as well as practiced alongside traditional spiritualities. When we say the Land is sacred, we are being spiritual and describing a spiritual relationship with Land, the same way when we say, "defend the sacred." Many things are sacred to Indigenous folx, and many acts and traditions are medicine, meaning that they are related to healing and unifying with the sacred. Here, Jeffrey and I talk religion, and we frame out that this tends to be associated with the political right wing, which neither of us relates to. It does, however, bring us into contact with uncomfortable questions and discussions, especially when it comes to right-wing Christians who expect the same from us:

VEREGGE: Yeah. And that's what blows my mind, is that the other aspect of negative social media is that I'm happy because we're seeing things, light shed on things. But you're also shedding light on people you knew and loved and thought you knew for years. And it's like, wow. I can't, I can't believe you just said that, I can't believe that you believe that and—

TEKOBBE: I don't know how to process that. I really don't, especially people I love and care about. I . . . Do you want to correct them, but how do you, how do you do it? That is just one of those things that really saddens me and frustrates me, is when I see people that I, that I cared about and saying things that I can't support.

Clearly, there is not a gray zone. No.

VEREGGE: Yeah, yeah, no. And that's, and then, and that's why I get, when I see people, and that's probably where I've lost people, is where I see them quoting the Bible or misquoting it. And they're—I let them know, that's not what that says. I did take theology courses, I did take apologetics courses, I do read my Bible. And so, I think about a lot of those sorts of things and I would hate to see a message, I hate seeing a message of love being used as a shield to hide people's own insecure private insecurities, and using it as a device for that.

TEKOBBE: I agree. 2020 has been a thing, yes. And so, I guess I don't want to make you out to be the role model. Like, it's really hard when you're a person of color, you're an Indigenous person. And people come along and they're like, what do the Indigenous people think? What are the Native Americans thinking? But I just, I don't like that. That's not what I'm trying to do. But I just want to know if you had any advice for other activists and artists doing this kind of work and they're engaged in social media and engaged in their own work. Did you have any advice?

Here, I bring up the year 2020 again, which we were talking about earlier in terms of the presidential race, the fires and other environmental disasters, and bigotry and racism on display across the country. I say again that I do not want to hold Jeffrey up as an example of all Indigenous artists. I do, however, think, particularly for those same artists starting out and who know of Jeffrey's work at the intersection of science fiction and fantasy and indigeneity, that Jeffrey has a lot to offer.

VEREGGE: Be true to yourself, is the first thing. Always know who you are and remember where you came from. That being said, also remember that where you learn, take what the very best from where you're from and use that in your life. Use that on your social media—don't take no for an answer. You be smart about things, you don't have to be drawn into every conversation every day, every minute, every hour, somebody is going to post something that's going to offend you. Something's offensive, something that's going to be wrong. You can spend time, what I've learned the last six months is, you can spend time arguing with that person, trying to bring them for the other side. But a lot of times they're there for whatever reason. They are set in that mindset. And you've just wasted all that time.

Know when to walk away.

Know when, like Kenny Rogers says, know when to hold them, know when to walk away. Know when to run. You know. It's one of those things, that is, be smart about your conversations, engagements and remember, everybody is watching. And it's things that you say nowadays, especially online, can be brought up whenever.

And even if you are stupid, you can be intoxicated, you can be whatever. Totally different person. Five years ago, something like, wow, they posted that. You see it all the time, somebody goes and digs through some of these posts and it's just like, wow, I can't believe they said that. Remember who we are, what you want to represent when you're sharing things. Share your work first. And your inspirations second, and just remember that.

Don't worry about negative comments. You're going to get negative and positive comments. You can take it with a grain of salt, that. But as long as you're happy doing what you love. Don't let those tear away at you or your craft. You can use it, you know, if it's a negative, if it's a critique that's negative in a way that will make you better as an artist first.

Yeah, take that. Take it and go with it. Thank them for it. And, but if it's just somebody just trying to be a troll, just walk away. That's all there is to it.

I tell Jeffrey again that I think about being critiqued from a settler-colonial framework in which my work is not intellectual enough, important enough, scholarly enough. Here I am thinking about how he has been through some rough criticism for one or two of his earlier projects, both from within Indian Country and from without. He intimated that the criticism from Indian Country was harder, but you learn. And I thought about that, about that my fellow Indigenous scholars might read this, and they might find something useful here in the experience of my writing and telling these research stories. They might wish I did things differently, or wrote from a different position, or used different language. I tell Jeffrey that I think it is all very hard.

VEREGGE: Yeah. And it's, yeah. It's totally, it's totally. Yeah. This time its interesting in that the work that I'm doing with Marvel right now. I got so much great press a couple of weeks ago, almost, I would say 99, 98 percent positive comments. Everything else—as opposed to when I first did *Red Wolf*,[6] I didn't get so many great comments, especially from Indian Country. Indian Country was actually kind of the opposite of, of what everybody else was. And some of those comments were downright horrible and mean. Some made me laugh, because I had heard those terms used before and it was, you know, it was rough. The good thing was that

I was making a lot of friends as a result of those, the way I handled the situation. I didn't—I could have gotten mean and angry with everybody, but instead I chose to take each of the most of the negative comments one by one, and be positive about it and be civil. And when one guy ended up becoming a really good friend of mine, and he was not nice at all. I mean, it took, took a string of things, and he kept saying he'd go back and look at my other posts and say, "OK, now, yeah, OK," This guy said, "You really surprised me. So, I didn't think that you'd be like this." Well, that's exactly how I am.

TEKOBBE: When was *Red Wolf*? *Red Wolf* was five years ago?, I don't remember exactly. I remember responding positively to it. I just—what was the gist of the negative?

VEREGGE: That it was, I was in the Northwest. I was an artist from the Northwest doing art for a character based in Middle America, Southwest, Southwestern Native, that I was consulting the book and not writing it. But the fact of the matter was, I had many experiences at the time writing scripts for comics, and Marvel had brought me in to read and make sure that it wasn't offensive to Native people and to add anything that could be added to give it value. But I think that a lot of people kept coming up with an old Fort Indian, because I'm working with Marvel and the enemy and it's like, no, you don't understand. Marvel is—they *were* bad. They did some things, they added to the stereotypes, they are trying to correct that now. And it's not . . . it's a process, and just be a part of that. Correcting that which has been bad, has been awesome, but that's the thing that we come up with, the sorts of things it is.

I tell Jeffrey here that I worry about the same thing. I worry that some of the Chahta might read this and think that I am too western academic or too liberal or I misrepresent their experiences with my experiences. I do not want to be the voice for my people, even though I am speaking to our identities and our struggles and telling some of our stories.

VEREGGE: Yes, exactly. And that's the thing, like, I don't even speak for my tribe. I speak for myself, you know, and that's what I have to remind people, that I'm not The Voice, I am a voice.

And here is the western contradiction that we have circled back to—how we carry our families, our ancestors, our relations, and also be our own person while they are also their own people. This is something that is hard to make a western audience understand. Because self has thick meaning, and we are ourselves, while those selves are composed of others, entangled with our lands, our histories, our cultures. This is what we mean by tribe, and this is what we mean by ourselves.

> VEREGGE: You know, it's just having fun, is basically what I'm doing. I know that there are projects that I'm working on now that don't even feel like I'm working. I feel guilty for working on *Star Trek* right now for IDW, and I've been a Trekkie for, since I was thirteen, and so, seeing—being able to finally do that on official capacity. I got to do a cover for the Museum of Pop Culture when they had a *Star Trek* exhibit there, and that was cool, but I didn't really count it because it was a big exclusive for their use, the museum gift shop, and not necessarily one that was on the rack. So being able to do the *Star Trek* series on, that will be on the shelves, has been such an awesome, awesome reward, that. And be able to do it in my own voice. That's the coolest thing, is that all these things that are happening really has been a result of just being who I am now, creating artwork that I understood and that I love and just try not to put in—trying not to be something that I'm not.
>
> Yeah, yeah, I've been very blessed to be able to do that.
>
> So it's been, is, you know, I'm trying to think if there's anything else that I, that the future I look forward to, each and more opportunities are coming, different kinds of opportunities that I've only dreamt about.
>
> I'm going to continue to push myself as far as I can, and hopefully, I want, like, my sons, my kids to be better artists, if that's what they want to do. But I'm not going to make it easy for them. I'm going to . . . I'm not going to just quit. I'm just going to keep pushing it. But I want them to be better, is my hope.

I hope that for them too. I am glad to have made this relationship, one that will persist after this book.

5
MazaCoin

Decolonizing a Colonial Fantasy

The Little People

A long time ago in ancient time, while the Choctaw Indians were living in Mississippi, the Choctaw legends say that certain supernatural beings or spirits lived near them.

These spirits, or "Little People," were known as Kowi Anukasha or "Forest Dwellers." They were about two or three feet tall. These pygmy beings lived deep in the thick forest; their homes were in caves hidden under large rocks.

When a boy child is two, three, or even four years old, he will often wander off into the woods, playing or chasing a small animal. When the little one is well out of sight from his home, "Kwanokasha," who is always on watch, seizes the boy and takes him away to his cave, his dwelling place. Many times his cave is far away and Kwanokasha and the little boy must travel a very long way, climbing many hills and crossing many streams. When they finally reach the cave, Kwanokasha takes him inside, where he is met by three other spirits, all very old with long white hair. The first one offers the boy a knife; the second one offers him a bunch of poisonous herbs; the third offers a bunch of herbs yielding good medicine.

If the child accepts the knife, he is certain to become a bad man and may even kill his friends. If he accepts the poisonous herbs, he will never be able to cure or help his people. But, if he accepts the good herbs, he is destined

https://doi.org/10.7330/9781646426478.c005

to become a great doctor and an important and influential man of his tribe and win the confidence of all his people.

When he accepts the good herbs, the three spirits will tell him the secrets of making medicines from herbs, roots, and barks of certain trees, and of treating and curing various fevers, pains, and other sicknesses. That is the reason the Little People take the boy child to their home in the wilderness, in order to train Indian doctors, transmitting to them the special curative powers and to train them in the manufacture of medicines. The child will remain with the spirits for three days, after which he is returned. He does not tell where he has been or what he has seen or heard. Not until he becomes a man will he make use of the knowledge gained from the spirits, and never will he reveal to others how it was acquired.

It is said among the Choctaws that few children wait to accept the offering of the good herbs from the third spirit, and that is why there are so few great doctors and others of influence among the Choctaws.

It is also said that the Little People are never seen by the common Choctaws. The Choctaw prophets and herb doctors, however, claim the power of seeing them and of holding communication with them.

During the darkest nights, in all kinds of weather, you can see a strange light wandering around in the woods. This light is the Indian doctor and his little helper looking for that special herb to treat and cure a very sick tribesman. (Edwards n.d.)

Introduction

This is the story of what happened when an Indigenous person employed an emergent digital technology as a possible approach to addressing Native American reservation economic precarity, political autonomy, and poverty. In short, an Oglala Lakota citizen named Payu Harris launched the first Indigenous cryptocurrency in the United States. The effort went through two iterations before it finally failed sometime in 2016. The reasons behind the failure are complex, including biased reporting, settler-colonial notions of both capitalism and Native American reservation life, tribal suspicion of economic technologies, and an inability by journalists to imagine Native Americans outside a colonial paradigm. In this chapter, I will introduce the concept of cryptocurrencies, I will describe Harris's efforts, I will perform a close reading of specific news coverage around this project, and in the process, I will describe and demonstrate that colonial history and imagination interfered with Harris telling a new economic story for his fellow citizens in Pine Ridge.

Payu Harris is a citizen of the Oglala Lakota living on the Pine Ridge Indian reservation in South Dakota. In February 2014, Harris launched a Bitcoin variant, also known as an "alt-currency" or an "altcoin," one that he named MazaCoin (pronounced "mah-zah"), that he believed would stimulate the local economy, provide a stable local currency, provide economic autonomy, and do important sovereignty work and identity work for the Oglala Lakota people at Pine Ridge (Browning 2014). Bitcoin is a cryptocurrency. Simply put, a cryptocurrency is a currency that is generated and circulated online. As a currency, it does not exist in physical form, rather it resides in blockchains on servers. Bitcoin is the most widely known, mined, and circulated cryptocurrency. However, there are thousands of Bitcoin variants. Only a small percentage of these altcoins are popular, leaving the rest to be considered unstable or perhaps "junk" currencies. Bitcoins and altcoins are mined as part of a network. The "miner"'s computer verifies transactions in the Bitcoin ledger. When enough transactions are recorded to equal a block in the blockchain, miners can potentially receive a fraction of a coin for their participation in the process of verifying the ledger. There is no guarantee of miners gaining fractions of coins, and miners often participate in a "pool" with other miners, pooling their processing power, to increase their odds of successful mining. I am oversimplifying this process, which requires a fair amount of technological knowhow to establish, as well as electricity and network resource consumption, and I am definitely glossing over the environmental impact, but this is the gist: in exchange for miners' participation in maintaining the ledger of network Bitcoin transactions, new Bitcoins are slowly generated. Those few new Bitcoins are divided up between the people whose processors worked on the block of entries in the journal. Lots of people are working on the journal, and there is only a small chance of earning new Bitcoins for one's participation. This is a reason to pool efforts to gain a greater share of processing resources, and many miners participate in collectives to combine their processing power and share returns. While the results of mining are sometimes mixed, bitcoins are eventually generated in the process, and over time will accumulate as proceeds. The general thinking is that it is possible for a stable, independent economic system to be built on cryptocurrency.

Harris began work on his altcoin project, and before long, the story of the Native American Bitcoin variant was making its way around the world in mainstream digital media. To digital journalists, the allure of the lone Indian living on Pine Ridge reservation, the site of perhaps the most research into Native American life, culture, and social issues in the United States, was apparently irresistible. Industry and mainstream digital journalists wrote themed stories of Harris as a warrior rising out of the past to take on the federal government once again. They wrote traveler's tales and drew on Indigenous history with little differentiation between colonial legend and historic fact. These stories and surrounding popular warrior imagery caused the interest in, and thus the value of, MazaCoin in cryptocurrency markets to rapidly rise. Harris was an instant celebrity in these cryptocurrency circles—until the backlash from tribal members and Indigenous publications destabilized the currency and caused the value to fall so low the initial project effort was defunct over the course of eighteen months.

Given that the project of colonialism is the exploitation of Indigenous resources, cultures, knowledges, and bodies, then the downfall of the Indigenous cryptocurrency project is perhaps unsurprising. In capitalism, there are always colonizers. Colonizers claim that which belongs to the "other," squeeze everything they can out of it, and then erase or blur over their colonizing practices, often leaving ruin behind. Once the digital journalists had everything of value from the story, they abandoned it to the controversy that was created. They abandoned Harris to navigate the fallout on his own.

It is important to note that I did not interview Harris for this project. I made my requests, but he did not respond. Online connections of his, those who appreciated the *First Monday* publication, tried to convince him to engage me, but he did not respond to them either. Here, in "strategic contemplation" (Royster and Kirsch 2012), I have spent a lot of time thinking about why Harris would not want to participate in an interview with me. At first blush, I think it is obvious that he has done a lot of press, and it has not benefited his project. In that light, how would yet another interview benefit Harris? Or perhaps more importantly to him, how would talking with me benefit his project? And in terms of the reciprocal nature of Native relationships, what do I have to offer Harris in reciprocity? The chance to tell his story again? For many reasons, this probably does not seem like a worthwhile outcome to Harris.

I also think that the history of research on the Pine Ridge reservation, some good and much objectifying, means that a scholar requesting access to a Pine Ridge citizen could be met with skepticism. Finally, and maybe most significantly, in the Indigenous concept of relations and relationships, I have no relations in common with Harris. He does not know me, and he does not know my people. I do not know his. We have no one in common to make introductions, to share experience or help us establish trust. And certainly, Harris has his own reasons. With these considerations, I think it is clear that there are sound explanations why Harris would not participate in an interview with me. In this chapter, then, I will not report beyond his words in the media or report anything specific about Harris. I will, instead, share and analyze the digital media storytelling about Harris, and how the digital press's racist, colonial, and capitalistic lenses helped derail Harris's project. Harris's personal story is and should be his own to tell if and when he chooses to.

Collecting data on cryptocurrencies, and on tribal interest in those cryptocurrencies, turned out to be a complex and messy process. Typically, when I research cultural stories and cultural rhetorics, I spend a lot of time reading periodicals and news sources, first more broadly, in such venues as the *Washington Post* or the *New York Times*, and then more industry-specific, like in financial periodicals or film industry publications, depending on the audience or community from which the culture I am looking at is constructed. I am looking for general public discourse around my topic, and I am also mining those news and culture pieces for more concrete sources: for example, a *New York Times* article might quote and cite an author of a book, and that book might have more complex information. If I can get several complex sources, I can build a more multidimensional picture of how the culture is operating and speaking. I am filling out the story. The initial problem I ran into when researching MazaCoin was bias, but not racial bias—that came later. The bias I first encountered was pro-Bitcoin marketing, because a number of websites reporting on cryptocurrency trends are cryptocurrency advocates and have an agenda to expand the circulation of Bitcoins. Their site advertising is also sourced directly from the cryptocurrency industry. I suppose that seems obvious, but I was looking for news stories about cryptocurrencies, and what I got were glorified promotional materials that made it difficult to get a sense of the cultural landscape beyond sentiments like "capitalism is good" and "private money is better than dealing with a central government treasury." As a result, most of the stories I found were in industry-specific websites, podcasts, and social media feeds. There were only a few scholarly sources on alt currencies in 2016. The majority of scholarly articles about cryptocurrencies were published from 2018 onward, and are in the fields of economics, finance, and, with less frequency, business. So, in general, my difficulty was that telling a story about cryptocurrencies was more complex than researching some other digital cultural movements, because in general, cryptocurrency stories do not find their way into the most major mainstream news streams.

Another difficulty with the research was that Native American stories also rarely make it into major mainstream media streams. There are essentially two common types of news stories around Native folx: (1) those written on occasions when Indigenous issues overlap federal interests, like when there is a court battle over water, and (2) the occasional special interest story that is told with a particular "take," usually with Natives as a cultural anachronism. Again, this is probably an obvious stereotype, what with the perception that Natives are isolated on reservations so that they rarely interact as a legal entity with other governments. This is a stereotype, because there are Native issues all around us with land management, natural resource management, child adoptions, voting rights, and more, as I have demonstrated in this book. But the obviousness in this observation is that Native issues are, without Native activism, largely invisible and are only made visible when their interests are in conflict with white interests or their culture is being particularly colorful, like in a regional special-interest announcement of a local powwow with hoop dancers and fry bread. Again, this is obvious. But what I am getting at is that mainstream media decides which stories have mainstream relevance, or are worthy of being reported, and that happens in the case of Native issues only when white viewpoints need reporting. Which is why 2014's MazaCoin caught my attention. Following the reporting in *Russia Today*, MazaCoin became a mainstream news story. From 2014 through 2016, MazaCoin, or simply Maza, as it is was called later in the project, and its backers struggled to find their footings and their identity among the modern gold rush that was the cryptocurrency marketplace at that time. These circumstances made a story of interest to large media outlets like *Newsweek*, *NPR*, and *Forbes*. The research presented challenges, then, as I sought information to build the research story of MazaCoin beyond a settler-colonial narrative.

Methodologies

This project applies Indigenous rhetorics, cultural rhetorics, digital rhetorics, and intersectional feminist research methods to attempt to engage the early efforts of Payu Harris, a member of the Oglala Lakota community, to advance the new cryptocurrency MazaCoin as a solution to his community's economic and social issues in 2014. First, I introduce Harris and his MazaCoin project, describing the project's rapid rise and fall. Then, I theorize decolonizing efforts as they intersect with the capitalism and neoliberalism of digital spaces. Next, I examine several key moments in the MazaCoin timeline through this theoretical framework. Finally, I consider the future of decolonizing efforts as they encounter neoliberal digital platforms and networks.

Here I argue that Indigenous digital activism is viewed by the media and the colonial networks as it always is. It is imbued by settler logics, and it can entrap peoples in the process. Then I will tell the story of Harris and MazaCoin, their rise in digital media, and the ultimate failure of Harris's project. I will conclude with a discussion of why this research matters for Indigenous and humanities researchers as well as researchers of the sociotechnical.

I begin this chapter, in the decolonizing and Indigenous rhetorics narrative practice of sharing stories to create the groundwork for accountability and solidarity, with the story of how I came to work on this project with John Carter McKnight (Bratta and Powell 2016; Dougherty 2016; King, Gubele, and Anderson 2015). John and I met as graduate students at Arizona State University (ASU), Tempe, Arizona, where we were sharing research space in James Paul Gee's Games and Impact lab at ASU. Over many key clicks and shared pots of hot tea, we learned we were both born and raised in the American West, with its wide-open desert landscapes and ranging and diverse American Indian reservations and territories—a shared identity with which we both feel deeply connected. We share similar alternative paths to academe, with my returning to graduate school after a career in the technology industry and John returning to graduate school after a corporate legal career. Both of our work histories intersect the financial sector, and our individual research interests both include networked technologies and social justice. We like cats. We are close friends.

In 2014, John was studying peer-to-peer finance in a postdoctoral research appointment at Lancaster University, United Kingdom, and I was completing my dissertation at ASU. In February 2014, John saw an article written by Danny Bradbury on CoinDesk.com[1] of a new Bitcoin variant, MazaCoin, launched by the "Traditional Lakota Nation" (Bradbury 2014). John sent the article's link to me, asking if I had seen any coverage of the unlikely story in North American or Native American media. I agreed it seemed unlikely and suggested we each look for additional online media references and then Skype to discuss our findings. At the center of our interest in the MazaCoin story were (1) an abiding concern for the representation of the Oglala Lakota people and culture by a media industry that tends to represent Indigenous peoples and communities as little more than poverty porn (Jensen 2013, 2014) for mass consumption, (2) a scholarly investment in the emerging alt-currencies as cultural identity work around the world, (3) the early potential for alt-finance practices and platforms to reach underserved communities, and (4) a deep skepticism about any rapid adoption of a technocultural project in the economically tenuous

Oglala Lakota territory. What followed was almost two years of tracking the media coverage of the rise, fall, and attempted resurrections of the MazaCoin project, coverage that in its language, images, and positionalities mirrors the conflicting and intersecting interests of cultures, identities, technology narratives, platforms, peoples, and histories.

We submitted our findings in a paper to the annual conference of the Association of Internet Researchers and were accepted. John took ill, so I presented our work to the conference in October 2015 in Phoenix, Arizona. My session was held in a conference venue that had meeting rooms named for Arizona Native American nations and which also included communal spaces constructed in homage to traditional Native American kiva architecture. These are relatively common features in Arizona, which is home to more than twenty sovereign Native American nations. A kiva is a circular or rectangular recessed floor that accommodates a group of people seated in a circular configuration to meet, socialize, engage in political conversation, and conduct ceremony. It was striking to present our research to an international audience gathered in such a place, in those conference rooms with tribal names. I stood at the front, showed my slides, and gave my talk to an audience that was seated in front of me. I would have preferred, with my decolonial research and methodologies, to step outside the room and use the kiva space, one where we could all look at each other and share in the knowledge without the hierarchy, but I felt like conference conduct at the time would not allow for me to take over communal spaces. In 2020, after five years of presenting Indigenous and cultural rhetorics at conferences, I feel like were I to be in this situation again, I would move our whole session out to the kiva where we could be collaborating and doing research as cultural work in the space as its design intended. Still, at the time, the contrast between my presentation of our methods and findings in a conference room and the appropriate venue for those findings right outside the door yet inaccessible stays with me.

For the conference, John returned to Arizona from the United Kingdom, and I from my new position at the University of Alabama in Tuscaloosa, Alabama. We acknowledge this conference brought us full circle: a return to the Southwest with its significant and visible Indigenous presence, and a return to our research partnership. We published our findings in the internet journal *First Monday* on October 3, 2016. John completed his postdoctoral appointment and accepted a position in Pennsylvania, while I continued with my research in Indigenous identity and digital activism at the University of Alabama. We, and our project, are grown from a specific space and time, and it is specifically this awareness of space and time that we wish to convey to our readers as they consider our analysis of the MazaCoin endeavor grounded both in our space and time and in Indigenous notions of space and time. For us, and for this project, the concept of time meant several different things—our working relationship separated by time zones, our research enabling us to spend time together, the blip of the timeline of our project in comparison with the hundreds of years of the colonial project. And also, we were working on Indigenous time in that we wanted to do good work and we wanted to take the time the project took. Yes, we had editorial deadlines, but if we could not write the project in a good way, we would not have wanted to finish it. This chapter is organized around linear time, but for John and I, the original research project unfolded within our relationship, with text messages and video calls, emojis, and the kinds of asymmetrical communication that happens when one of you lives in the American South and the other lives in the United Kingdom. We tended to move back and forth between research questions and data, and between discussion and contemplation. Our practice was decolonial, but this chapter is organized along linear time, in the way that scholarship is typically read; however, I disrupt this timeline at points, not to bring confusion but to fill out the story.

Linda Tuhiwai Smith writes of the distinctions between Indigenous place and time and western notions of place and time in her text *Decolonizing Methodologies* (2012). She explains that Indigenous place and space is connected to people and relationships. Spaces and places are meaningful in the connections between the people and the land. It is western practice to divide up spaces with linear borders and assign them names as a means of conquering the space. It is western practice to create hierarchy and taxonomy. It is western practice to make hard divisions between work, recreation, family, and resting times. She notes how the colonial perceptions of Indigenous "laziness" are related to these notions, as colonist observers were baffled by the lack of division between these various activities and times. Distance as a concept serves western objectification. It calls for objective distance. It is the abstraction of words, ideas, stories, relationships, and histories. Notions of time, space, and the divisions between these are also closely tied to capitalism. Production must be completed on a specific and efficient timeline. Cryptocurrencies are a kind of hypercapitalism. This definitional work I am doing here is important, because John and I were not driven by western concepts of appropriate time and place, nor by expected efficiencies and production. We worked organically across geographic and time zone boundaries, and we researched and wrote in spaces in the conversation that passed between us as part of our greater relationship. This distinction is important, I believe, because we were consciously engaging in decolonial practice in our scholarship and our relationship, but the linear organization of this story erases some of that decolonial work. I draw your attention to it in hopes you will keep it in mind as you read the events described here.

MazaCoin: A Cultural Convergence

During the two years of MazaCoin's launch and eventual shuttering and then rebranding and relaunching, I followed as many news stories as I could find. In 2019, MazaCoin was rebranded as Maza, but it has since collapsed again, for possibly the last time. Officially launched on the altcoin exchanges on February 20, 2014, MazaCoin first came to John's and my attention from the February 6, 2014, *Coindesk* article "MazaCoin Aims to Be Sovereign Altcoin for Native Americans" by Danny Bradbury. *Coindesk* is a digital currencies news website. The article introduces the entrepreneur behind MazaCoin, Payu Harris, as a member of the Oglala Lakota Nation. In the article, Harris is described as launching MazaCoin under an umbrella project name, Bitcoin Oyate Project, or BTC Oyate,[2] and he is quoted as describing the purpose of the altcoin as offering fiscal autonomy to Indigenous peoples: "An independent crypto currency would eliminate the State/Federal ability to freeze accounts and tamper with lawful tax revenues." Harris further explains that BTC Oyate is designed to specifically benefit the Oglala Lakota of the Pine Ridge Indian reservation, where he resides, but he also envisions cryptocurrencies as being beneficial to other tribal groups: "If a tribe forms an update or brings up a direction they feel is important, representatives for each tribal Economic Development Administration or crypto currency development directors can come together and vote on adoption by other regions or tribal governments." Harris envisioned an expansive Bitcoin that would hold half the proceeds in tribal trust, and the other half would sustain the Oglala Lakota economy (Ramos 2014). He conveyed to Bradbury that he believed the project would see wide adoption by tribes across the United States.

While Harris describes the "crushing poverty" of the Oglala Lakota from his own experience, Bradbury applies United States census data to describe the aggregate poverty of Native American households as twice that of the national average of non-Native households. His assumption that there is some kind of parity between the conditions of Native communities and non-Native communities compresses the distinct differences between Indigenous peoples and their own tribes and nations, and the settlers who colonize them. Here, Bradbury subsumes the Native American experience into homogenous generalizations about poverty as the economic context for the Oglala Lakota of the Pine Ridge Indian reservation Harris calls home. This complex economic context is specific and emerges from the Oglala Lakota history of relocation and colonization, rather than being something that is possible to generalize from aggregated census data. Bradbury's attempts to validate Harris's experience against national data serves to colonize Harris's narrative and erase the colonial impacts of capitalism and imperialism on the Oglala Lakota. Harris sees and feels and experiences this poverty in his daily life. Bradbury frames the experience out in generalized census data.

On February 27, 2014, *Forbes* ran an article by Jasper Hamill, "The Battle of Little Bitcoin: Native American Tribe Launches Its Own Cryptocurrency." Hamill quotes Harris as inspired by gazing over the site of the Battle of Little Bighorn: "Suddenly the story of Custer's Last Stand wasn't just words on a page but something deeply personal. I looked at how things were for the tribe now and suddenly had an idea about how we might fix it." Hamill also describes Harris as an Oglala "chief" and declares that MazaCoin "has now been officially adopted as the national currency of the Lakota Nation." Hamill briefly references some Native confusion around how the new currency could help the Oglala Lakota, and then he reports that the FBI has contacted Harris to warn him about the illegality of cryptocurrencies. Hamill bookends his piece with martial metaphors, calling Harris the "son of the once-mighty Oglala Lakota Tribe," and linking the MazaCoin initiative to historic Oglala Lakota battles with the United States military. Here, rather than generalizing the Native American experience, the journalist appropriates and romanticizes the "once mighty" Oglala Lakota, spinning a tale of scrappy uprisings and the triumph of the noble few against the overwhelming federal forces.

While the whole piece has moments that are problematic, one particularly egregious move here is with the editors choosing to repurpose Harris's own take on Custer's Last Stand as an attention-seeking headline, "The Battle of Little Bitcoin." For Harris, his own storytelling efforts here are about inserting and employing financial technology on the historically long road of tribal financial interdependence with the federal government. There are many misconceptions that circulate about and around Native Americans, as many people believe Natives receive government payouts for housing, education, healthcare, and the like. The truth is a lot more complicated, with tribal funding heavily entangled with agencies and services. I argue here that because of Harris's lifetime of firsthand experience, when he talks about righting the wrongs of the past and being inspired to take action through a launch of the tribe's own currency, he is referencing much thicker meanings than simply a linear savior story. Contrary to the headline, Harris is not calling himself Crazy Horse, he is situating his story into a long history of federal/Oglala Lakota conflict.

At the beginning of March 2014, Nick Spanos, an entrepreneur and the director of the New York Bitcoin Center, flew Harris out to ring the opening bell at the Center and meet with the financial media, altcoin press, and potential MazaCoin investors. Between February and March 2014, Harris completed many interviews and received global press and financial industry attention. MazaCoin was trading briskly on the exchanges, buoyed by financial media interest and the image of a small, upstart Native American resistance against the mighty federal government, which dovetailed well with capitalist and libertarian narratives of self-determination, neoliberal bootstrapping, and anti-regulatory sentiments that are closely interwoven in altcoin ideologies. These narratives were seemingly at odds with Harris's vision and project. Harris wanted a collective solution; he wanted to raise money to hold in trust for members of his tribe. He wanted to shore up the Oglala Lakota economy by making the MazaCoin currency the primary currency for those local businesses. He wanted to improve conditions across the board for everyone on Pine Ridge, allocating shared resources as needed, in contradiction to libertarian ideologies of self-interest (Boaz 2019) and neoliberal ideologies of the value of individualized success. While it is certainly possible that Harris believed he could improve his own financial position, that was not among his stated goals. The bloggers, writers, and reporters were likely only able to focus on a single story because of the need to describe and classify Harris—but that does not seem to be the story that Harris was trying to tell.

Trouble began when on March 3, 2014, MazaCoin was reported on to an audience primarily of Native American peoples and those interested in Native American issues. An article by Alysa Landry ran in *Indian Country Today*, a Native American news website that covers stories of Indigenous interests across the United States and Canada, "9 Questions Surrounding MazaCoin, the Lakota CryptoCurrency: Answered." In her piece, Landry describes the Pine Ridge reservation as encompassing the second poorest county in the United States, which is a clearer representation of the economic conditions at Pine Ridge than many other media stories, because she narrows her focus to Pine Ridge, not an aggregate of financial data across all tribes. Landry also describes Harris as "a 38-year-old dad who once operated a video store on the reservation" and identifies his affiliation as "Northern Cheyenne." Landry contacted John, and she quotes him describing the speculative and unstable nature of cryptocurrencies and advising against them for tribal people whose interests are in protecting and cultivating their current assets to benefit the greater community. Between cautionary statements and alternative suggestions by John, Landry sandwiches a statement from Harris that he "believes as many as 50 percent of [reservation] merchants will buy into the system within the first 12 months" and that he hopes to see cryptocurrencies as a standard for all Indigenous communities within ten years. This positioning of analysis of a white scholar suggesting that asset preservation and cryptocurrencies are incompatible, set against Harris's discussion of his own plans for the altcoin, created an artificial opposition of white authority and Native experience. This certainly was not John's intention; however, this again is how whiteness operates, with internalized colonialism applied to an investigation by a Native press. This journalistic shift toward doubt appeared to reflect the growing unease surrounding MazaCoin on Pine Ridge. However, it was also a shocking example of the difficulties of doing decolonial scholarship, given the report of a white scholar was used to validate the viability of an Indigenous project, rather than accept Harris's narrative of his own altcoin. John and I were both disappointed by this outcome, and it is likely another reason Harris did not want to speak with me.

On March 7, 2014, *Native Sun News* ran "Oglala Sioux Tribe Surprised by MazaCoin Plan" by Brandon Ecoffey, its managing editor. Ecoffey subtitles his article "Man Claims OST Has Launched Own Currency; Council and President Taken by Surprise." OST is an abbreviation for Oglala Sioux Tribe. Ecoffey opens his story with "the Oglala Sioux Tribe supposedly launched its own national currency. However, no one bothered to inform Tribal President Bryan Brewer or the Oglala Sioux Council." Ecoffey references the *Forbes* article's claims that Payu Harris is a chief and a member of the Oglala at Pine Ridge, yet he argues that Harris's name does not appear anywhere on tribal rolls. Here, Ecoffey relies on colonial accounting, those rolls, of who is Indigenous and who is not, to supposedly invalidate Harris's authenticity. This use of tribal rolls and registries to verify enrollment and therefore account for the distribution of resources is a common and highly contested practice with, as I discussed in the introduction of this text, many people being left out of claims of Indigenous identity because their families were not included in the records. There is the possibility that Harris is not Oglala Lakota, but he has lived and owned businesses there. Either way, his membership among the OST is an internal, administrative matter, while Harris's narrative is an identity matter. The comments thread on the *Native Sun News* article had the BTC Oyate Initiative responding to the reporter's questions, contradicting that Harris ever claimed to *Forbes* he was a chief or that MazaCoin had been officially adopted by the OST. Dana Lone Hill,[3] listed as a contributor at *Last Real Indians*, an independent Indigenous media website that runs stories of events and issues in Native American communities in the United States and Canada, responded to the BTC Oyate Initiative by leaving a comment in the comments thread stating that Harris lacks credibility because he is available to the mainstream media for interviews but not to the *Native Sun News*, and she asks, "How could anyone ever trust money you can not see?" There appears to be at least three separate questions of trust here. First, Harris was, according to *Native Sun News*, not trustworthy, because Harris's name does not appear in tribal roles—is he a real Oglala Lakota? Next, could Harris be trustworthy in the media, because his identity is misrepresented—how did *Forbes* make the determination that Harris was a "chief"? Third, is the currency trustworthy, given its technological and colonial origins? Lone Hill was not the only skeptic, as stories and comments spread through forums and in the Pine Ridge community that Harris had, at best, overstepped and in he which he was, at worst, a con artist (Consunji and Engel 2014).

By the end of March 2014, like many other overhyped financial instruments, MazaCoin had collapsed to near worthlessness on the Bitcoin exchanges. A few weeks after that, Harris parted ways with his coding partner in the initiative, and Spanos declared the project all but dead. Supporters of MazaCoin then moved to a Reddit forum, where they attempt to disseminate information and assist Harris as grassroots cryptocurrency supporters who believed they had found something unique in the Native American ethos of MazaCoin. In September 2014, *Mashable*'s Bianca Consunji and Evan Engel described Harris as another failed cryptocurrency launcher, "unfailingly optimistic, technologically obsessed, and supremely self-assured." At some point in 2015, MazaCoin and the BTC Oyate Initiative website, Facebook page, and Twitter accounts were deleted. The effort is reorganized as simply Maza on the new website, and it is more broadly characterized as an Indigenous-initiated cryptocurrency with mass appeal. There is minimal press about the new Maza, and the hype has disappeared (Williams 2018). The open-source code is moved to GitHub, but there is scant public information available about the development team or the project's sponsors, which is either sympathetic or suspicious, depending on the positionality of the observer. I tend to lean toward a sympathetic position. Across media characterizations, Harris is a deeply invested member of the Pine Ridge community, a snake oil pitchman, a collectivist, a technolibertarian, an insider, an outsider, a traditionalist, an innovator, a warrior, or an opportunist. He and his project are maybe all or none of these things, but MazaCoin and its founder offer a unique opportunity to discuss current Indigenous decolonizing and sovereignty movements as they intersect digital financial spaces and monetized platforms.

Theorizing the Digital Decolonizing Efforts

Indigenous scholars remind us that research as it involves Indigenous peoples has a racist history and continues to be troubled ground: "Research 'through imperial eyes' describes an approach which assumes that western ideas about the most fundamental things are the only ideas possible to hold, certainly the only rational ideas, and the only ideas which can make sense of the world, of reality, of social life and of human beings. It is an approach to indigenous peoples which still conveys a sense of innate superiority and an overabundance of desire to bring progress into the lives of indigenous peoples" (L. Smith 2012).

If the goal of Indigenous scholarship is to address political and social inequities and to heal and decolonize communities, then I must not usurp the autonomy of the Oglala Lakota, Payu Harris, or other Indigenous voices as I attempt to analyze and interpret the reception and collapse of the MazaCoin project (Suzack et al. 2010). I am deeply mindful that the Oglala Lakota of Pine Ridge are perhaps the most studied of Native American communities, and of deep harm to the Oglala Lakota perpetuated by the false and racist thesis that the poverty and social ills of Pine Ridge boil down to the confinement of a warrior tradition without weapons or battles left to fight (Deloria 1988). My task, then, is not to rely on colonial hierarchies for organizing information but to accurately represent what John and I collected and be transparent about how I frame and relay that data. Thus, this is not a study of the Oglala Lakota or their economic conditions or decisions. I do not view Payu Harris as a text or passive subject; again, Harris's narrative is his own. Rather, my interest is in how Harris and MazaCoin are constructed by media representations and choices, which are surely influenced by the cultural beliefs and stereotypes that surround Native American peoples.

Financial technologies create classes of users and nonusers (e.g., the "unbanked"), which are true classes, not mere categories. While, arguably, all technologies are "neither good nor bad nor . . . neutral" (Kranzberg 1986), financial technologies reify social relations and attitudes around money, which in most western cultures is closely coupled with notions of moral, as well as economic, wealth and poverty. Cryptocurrencies are strongly associated with the far right as being against centralized banking and oversight of marketplaces. They are also the domain of the technological elite. As Adam Greenfield writes in *Radical Technologies* (2017, 117):

> For all the hype around Bitcoin, it is clear that in its design, important questions about human interactions, collaboration and conviviality are being legislated at the level of technological infrastructure. Its appearance in the world economy gives disproportionately great power to those individuals and institutions that understand how it does what it does, and are best able to operationalize that understanding. At present, only a very tiny number of people truly grasp how Bitcoin and its underlying technologies work to create and mediate the transmission of value.

The design of these technologies, from cash to publicly traded corporate securities to a range of "alt-finance" innovations, inscribes and communicates messages about social as well as financial value: consider the difference between paying for a purchase with rumpled dollar bills, a black American Express card, or Apple Pay on an iPhone. People who have more value-inscribed assets, like the fabled black American Express card, are viewed as more economically successful (and in a western, capitalist paradigm, more moral, as work is tied to assessments of character and morality) and thus have a higher "class" status than those with lower value—inscribed currency, like the crumpled dollar bills.

Although Bitcoin is a highly technical financial instrument, and therefore one marked as higher in the hierarchy of social value in relation to other currencies, its narrative is highly individualistic, if often pseudonymous ("Satoshi Nakamoto," the inventor of the blockchain underlying Bitcoin and its variants, continues to conceal any identifiable personal details), rooted in libertarian, even Randian, notions of individual value creators in opposition to "corrupt," "outdated" governmental institutions, particularly in the realm of finance and "fiat money." Cryptocurrencies like Bitcoin and MazaCoin require significant sophistication in both computer science and finance to use, and extensive computational and electrical power to maintain. This design is neither accident nor coincidence: despite the claims of universal uptake by cryptocurrency advocates, and similar claims that "anyone" can make a Bitcoin, the cryptocurrency's design and ideology both mandate a distributed technological elite beyond the control of governments, a notion dating back to John Perry Barlow's 1996 "Declaration of Independence of Cyberspace:"

> Governments of the Industrial World, you weary giants of flesh and steel, I come from Cyberspace, the new home of Mind. On behalf of the future, I ask you of the past to leave us alone. You are not welcome among us. You have no sovereignty where we gather. . . .
>
> You have not engaged in our great and gathering conversation, nor did you create the wealth of our marketplaces. . . .
>
> Your increasingly obsolete information industries would perpetuate themselves by proposing laws, in American and elsewhere, that claim to own speech itself throughout the world. These laws would declare ideas to be another industrial product, no more noble than pig iron. In our world, whatever the human mind may create can be reproduced and distributed infinitely at no cost. The global conveyance of thought no longer requires your factories to accomplish.

The full document is worth a read, especially if you are fuzzy on your knowledge of the early days of the internet and the ideologies that it was founded on. However, it is clear in the excerpts I have included here that cyberspace was founded on the idea that internet transactions are beyond the scope of the government to regulate. In the case of MazaCoin, the tribal governments would be adopting the altcoin in order to have an "official" cryptocurrecy of their community.

The early discussion of MazaCoin seemed to conflate these two narratives: spontaneous action of non-elites and a high-profile individual advocate, anti-government rhetoric and an apparent governmental endorsement, sophisticated hardware and software for a region noted for its poverty. John's analysis of the sociotechnical assemblage of peer-to-peer lending in the UK, which had taken a significantly different turn from its US counterparts, suggested two outcomes of the project: either abandoning a populist assemblage in favor of a product aimed at elites, as happened with peer-to-peer finance in the US, or attempts to modify both the technology and its associated narratives to appeal to a non-elite audience, as was underway in John's peer-to-peer finance research in the UK. In my view of John's analysis, some of Harris's narrative of his history and the history of his community are rhetorical moves to redefine the cryptocurrency's associated technologies and narratives as one that was less elitist and more communal. In other words, Harris was composing a new story for Bitcoin, attempting to tell a story into being, a story of a thicker meaning for cryptocurrencies. In general, Indigenous North Americans are not strongly capitalist, in part because of its association with colonialism, in part because land was collectively owned until recent memory, and in part because communalism is retained in some of their cultural identities. Harris's story would carve out this space for broader meanings and usages of Bitcoin, combining communalism and independence from federal financial agencies.

MazaCoin: A Media and Digital Forum Construct

The MazaCoin site was offline for several years. It is now back up, and I am unsure when this occurred, only that it was done without media fanfare. MazaCoin has been a challenging endeavor that saw itself through multiple phases of participants and project goals. For example, for a time during Harris's project in 2015, MazaCoin was rebranded as simply "Maza," and the Bitcoin variant was described as follows: "Maza is like a new kind of currency for a new economy. It is being used by a social movement that has already built a new economy in the midst of the old. It is an economy focused on local community, inspired by ancient traditions of the Lakota people and implemented using innovative economic ideas that were developed over generations. The Maza economy is more sustainable, more resilient, more prosperous, more accessible and more fair than the old economy. Welcome to a new age" and "Maza: Welcome to the new (r)evolution" ("Maza" 2016). The website is HTML5 and built on an open-source template, "Spatial," authored by Templated, which describes spatial as "a simple, spacious, minimalistic design topped off with a large cover image" ("Spatial" 2016). The new website features images of Oglala Lakota in ceremonial dress and a clear description of Maza as "a decentralized payment system and distributed currency exchange," along with a discussion of the blockchain, ledger, and open-source code, which is available on GitHub (maza-online.com).

This glossier web presence replaced the original websites and Twitter accounts for the project, launched under the moniker BTC Oyate and its more homegrown-looking logo of a red, black, yellow, and white medicine wheel. Unlike the original Twitter account, there is no reference to the "National Currency of the Oglala Lakota Nation," and unlike the original website, there are no images of Oglala Lakota warriors and their iconic horses ("Lakota Nation" 2014). There are no specific declarations of sovereignty and Oglala

Lakota prosperity. The social media presence here has been polished, and the Oglala Lakota historical signifiers have been removed so that it appears the audience is no longer the Pine Ridge Oglala Lakota but an alt-currency audience looking for professionalism, stability, and investment opportunities. While the image of ceremonially dressed Oglala Lakota is a clear statement of Native American culture, it is a mainstream statement familiar to a greater public discourse that knows about powwows and fancy dancers but not specific tribal identities, histories, or the places where these converge. This is Native American culture for the masses (Gilyard 1999). Perhaps this is an attempt to sell an audience what they expect to see when they visit a site that is about cryptocurrency first and Native Americans second.

However, this newer, more glossy web presence is not the beginning of the positioning of MazaCoin, Harris, or the Oglala Lakota as consumables. In the February 2014 *Coindesk* article, Bradbury frames Harris's rhetoric of soverignty within journalistic objectivity, but that objectivity also undermines Harris's own agency. In recounting that one of Harris's goals is to give the Oglala Lakota fiscal autonomy from the federal government, Bradbury writes, "'One of the favorite tactics used by the Dep of Interior and the local State Government is to threaten to freeze bank acounts if the tribe takes a position that could challenge a State or federal interest,' [Harris] alleged." Bradbury follows with "Harris argued that the tactic has been used several times before against tribes operating casinos. 'An independent crypto currency would eliminate the State/Federal ability to freeze accounts and tamper with lawful tax revenues,' [Harris] said." Here Bradbury uses "alleged" and "argued" to describe Harris's recounting of Native American conflicts with federal and state authorities. Rhetorically, would Native American communities require an initiative that might bring them more fiscal autonomy if their autonomy was not historically threatened? Or, with the common knowledge that Native Americans were relocated by force from their ancestral homes, their property and assets seized by the federal authorities, is there no general public awareness that Native Americans have a history of less than complete fiscal autonomy? Clearly, there is a historical record to affirm a desire for Native American autonomy in general. However, Harris makes a specific statement that federal and state authorities coerce tribal compliance by threatening tribal assets. This is Harris's account of his own understanding of his own history. Bradbury is not obligated as a journalist to accept Harris's accounts of his own history wholesale; however, it is imperialistic to frame those accounts as arguments and allegations when they could just as easily be framed as anecdotal or tribal discourse (King, Gubele, and Anderson 2015).

Other media accounts decline to frame MazaCoin as a Native American story, in favor of creating a consumable product for audiences. The March 3, 2014, *RT News* article "Lakota Nation Adopts Mazacoin Crypto-Currency as Legal Tender" opens with the declaration that "a Bitcoin spinoff known as MazaCoin has been adopted by a confederation of seven Native American tribes as their national currency. The Oglala Lakota Nation has expressed hope that it will draw the tribes out of poverty." The article then drops in the non sequitur that "members of the tribe included Sitting Bull and Crazy Horse—famous for their role in the Battle of Little Bighorn, during which they were key in securing a victory over General George Armstrong Custer." John, whose scholarly interests include Russian popular media, came across this *Russia Today*[4] piece and sent it to me, asking specifically about the reference to the "seven" tribes. It is an interesting reference, because it does not seem to originate in the American media stories about MazaCoin. The Oglala Lakota are one of seven related bands of Lakota, commonly referred to in American discourse as "Sioux tribes." Equally attention-catching is the reference to the historically famous, and often romanticized, Lakota chiefs Sitting Bull and Crazy Horse, as well as the defeat of General Custer at the Battle of Little Bighorn. Together, these facts read like Wikipedia notes to the international reader, who might be familiar with the idea of Indians and the popular image of Sitting Bull but not the complex history or interrelatedness of the Lakota.

The popular image of the Lakota carries through many of the MazaCoin stories. Jasper Hamill for *Forbes* invokes the warriors-without-a-battle trope of the Lakota with his "The Battle of Little Bitcoin: Native American Tribe Launches Its Own Cryptocurrency." Perhaps an editor thought that play on words clever, but it also reinforces the nineteenth- and twentieth-century stereotype of self-inflicted Lakota poverty as the warriors refuse to change with the times, still frozen in their tidy dioramic battle with the American army (Deloria 1988). Hamill reinforces his metaphor with phrases like "son of the once-mighty Oglala Lakota Tribe" and "this time the war wouldn't be fought with arrows or bullets, but with QR codes and cryptography." Additional Hamill warrior references include "opening skirmishes," "great many battles to be fought," and "this war is just getting started." In only one of his quotes does Harris mention this history: when he states about Pine Ridge that his "family fought and died on this soil." In all other cases, Harris talks in language of communalism: "how we might fix [Pine Ridge poverty] and referring to MazaCoin as the 'new buffalo'" as a shared Oglala Lakota resource. Harris speaks of leadership, community, children, and prosperity for the Oglala Lakota. Hamill, in writing a fictional Indian war story, writes over Harris's self-reported account. He replaces Harris's language with his own, and he organizes his story around a colonial narrative of conquest (Sprague 2005). Payu Harris's account in his own words is rendered irrelevant in a western narrative, but it is hard to determine entirely the source here. MazaCoin is a capitalist initiative of a romanticized warrior people. Which narrative is more compelling to the audience of *Forbes*? The capitalism or the conquest?

As I mentioned in the introduction of this book, an early editor of the content this chapter is based on asked if I thought Payu Harris brought some of this on himself with his discussion of the history of his people and his positioning himself as a small start-up against the federal government. Is he responsible, in part, for the press's need to fall into stereotypes and traveler's tales? I think the answer to that question is complex. As I stated earlier, I think Harris's discussions of the Battle of Little Bighorn, and his references to his proud people, who have never ceded their sovereignty to the United States government, are rhetorical strategies to help rehabilitate his altcoin from the highly capitalist and highly individualist narratives of cryptocurrencies. I think this because Harris recenters his interview responses on what the altcoin can do for his people and their economy. He mentions not once about accumulating wealth for himself. I also think, given the online journalists' tendency to collapse descriptions of Harris and his project into nineteenth-century tropes of the American Indian as either warriors without battles or peoples without territorial homes, that those writers would have written their stories this way regardless of what Harris said himself. I think journalists cannot appropriate someone's narrative identity-work and then reflect it back on their subject when the story falls apart.

Conclusions

As of 2019, there are two primary MazaCoin presences on the internet, the glossy Maza website and the MazaCoin Reddit at http://www.reddit.com/r/MazaCoin/. The Maza site offers an investor experience, while the Reddit remains a forum for a few cryptocurrency enthusiasts who discuss "shamanic economics" and hold a kind of paternalistic interest in Native American matters. For example, one thread suggests using the "mazachain" to repurchase ancestral lands. The notion of purchasing lands that were seized from the

Lakota is not a cause of the people who are still in conflict with the US government over treaty violations. And it is not a cause of a people for whom the significance of the spaces is not in its measurable acreage or title but in its centrality to the unbroken cultural memory of the Lakota. You cannot repurchase what you never sold, and you cannot buy back that which was forcibly erased. The Lakota are not postcolonial, because for the Lakota, and other Indigenous peoples, the colonizers and their hegemony remain.

However, it seems unsurprising that the investor narrative and the paternalistic narratives have emerged from the original story of Payu Harris's MazaCoin, given the intersection of popular discourse, digital media, and affective capitalism. Given that Native Americans are still widely viewed as the nineteenth-century "Indian Problem," it is a simple matter to subsume their voices and overwrite their stories. While the underlying technologies of peer-to-peer finance could enable a system for both corporate elites in the US and middle-class investors in the UK, arguably the sociotechnical assemblage of a cryptocurrency and the self-expressed economic and social needs and values of the Oglala Lakota as its intended beneficiaries were simply incommensurable.

As a form of digital resistance to colonialism, I will continue to trace the efforts at alt-currencies by other Indigenous peoples. I continue to theorize that alt-currencies have potential for important cultural identity work, and that alt-financials have the potential to support the efforts of those traditionally underserved by the global financial sectors. Yet I remain aware of the tendency to colonize and repurpose the initiatives of minority groups, and I hope our scholarship in these areas serves as a potential antidote and a call for justice.

Conclusion

The Owl Woman
Some children were playing outside their house when they saw someone coming down the path toward their village.

"Who is that?" one child asked.

"I don't know. I've never seen someone like that before," another replied.

"Let's go and see who it is!" a third said, and so off they went to greet the stranger.

However, the children had not gone far when they saw that the stranger was a very old woman. Her body was bent with age, and her hair was white. In her hands, she carried a basket with a lid. She was so old and so bent that the children became frightened and ran back home. But soon, curiosity overcame them, and they went out to greet the old woman, who had arrived in their village at that time.

"Don't be afraid of me," the woman said. "After all, I am your great-great-great-grandmother! You have never seen me because I live very far away. Even your mother has never seen me! But maybe you can go and fetch your mother and tell her that I have come to visit."

The children did as the old woman bid them, and soon their mother had helped the old woman into the house and given her a deerskin to sit on. The mother and her children prepared a meal and gave it to the woman.

When the woman was done eating, she asked the children, "Tell me about your father. When he is home, where does he sleep?"

The children pointed out the place where their father slept.

That night, when the family was asleep, the old woman went to the place where the father lay and cut off his head. She put the head in her basket and covered the body with a blanket. Then she crept silently out of the house.

In the morning, the mother woke and started about the business of the day. She saw that her husband was still in bed, which was strange because he usually was the first one up.

"Are you ill, husband? Why are you still in bed?" she asked.

When her husband did not answer, the wife pulled off the blanket. She was horrified to find her husband's headless body beneath it.

Meanwhile, the old woman was hastening down the path away from the village, carrying the basket with the man's head in it. After a time, she came across a bear.

"Good morning," the bear said. "What do you have in your basket?"

"Oh, it is something very dangerous indeed. If I show it to you, you will instantly become blind. That's how bad it is," she replied.

The bear was alarmed by this and asked no further questions and went on his way.

The woman continued down the road until she met a deer. The deer also asked what was in the basket, and when the woman warned him that he would go blind if he saw it, the deer asked no further questions and went on his way.

All throughout the day, the woman met animals along the path. They all asked the same question, and she gave them all the same answer. The animals all left the woman alone after she answered them, until she came across two wildcats.

"Good day, old woman," the wildcats said. "That's a fine basket you have there. Can we see inside it?"

"Oh, no. I can't show this to anyone. Whoever sees what I have in this basket will go blind instantly," she replied.

"That doesn't matter to us at all," the first wildcat said as he tore the basket out of the woman's hands and lifted the lid.

When the wildcat saw what was inside the basket, he showed it to his friend. Both wildcats were horrified and angered by what they saw.

"We've heard of you, old woman," the first wildcat said.

"Yes, indeed we have," the second said. "You killed some of our friends. Now we shall avenge them."

Then the wildcats leaped at the old woman and took her captive.

> While the first wildcat held the old woman, the second went looking for something to use as a weapon to kill her with.
>
> Once the second wildcat was out of earshot, the old woman said to the first, "You know, if you really want to kill me, you should use that tree branch over there. It looks quite solid and probably would make a fine club. Also, it is good luck to kill me. Why wait for your friend to come back and let him have all the luck? You should do it yourself. I think you deserve it more, anyway."
>
> The first wildcat wanted that good luck for himself, so he let the woman go and went to pick up the tree branch. However, when he returned, the woman was gone, for she had turned into an owl and flown far away. (Clayton 2020, 19)

My mother was afraid of owls. I went through my entire childhood thinking about wise old owls, such as Woodsy Owl from the United States Forest Service, who told us to "give a hoot, don't pollute," and other white-people notions of good owls hoot-hooting from trees while wearing scholarly neckties and glasses. I listened to them calling out while my family and I were camping in the Arizona forests. I remember that she did not like the sound, but I did not know my mother was afraid of them until she was hospitalized and her room was on the top floor of the hospital. She could look out her window and see other areas on the roof. The hospital was using fake plastic owls mounted on the roof to scare away pigeons. My mother could see one of the fake owls from her bed, and she told me the owl was a harbinger of death. I cannot recall if we moved her bed, or if we had her moved to another room. I do remember that we arranged it so that she did not have to see the owl when she looked out her window. It was only during the process of writing this book that I came across this Choctaw story of the Owl Woman. I wonder if she knew it and thought it was too horrible to tell children, or if she did not know the story and had simply been told to fear owls by her mother, my grandmother. Sometimes I feel as if my memories know my mother only in fragments, and in some ways, this is a common experience for those with generational trauma where darkness is not always relayed in words, but it remains in shadows over and around families, both embodied and in spirit.

To write this conclusion, I need to address my research questions and findings. I also want to touch on what I have learned about Indigenous peoples' use of digital media since I began this project, as well as the scope of research I would like to do moving forward. However, from the questioning of identity (my own and others') where I began this project, I ended somewhere different, and I will begin this conclusion by answering the question of why I ended in a place that was not the place I began—how my questions expanded and perhaps mean something different now. And how, while the questions grew, I grew. And then I will attend to all the things a conclusion does, looking at the questions, looking back at the work, looking at the significant findings, and making clear a few takeaways about the research.

The first thing I want to say is that I cannot, from where I am standing, see the end of the COVID-19 pandemic, and this has dramatically impacted my ability to complete this project. As I write this, we are in our third academic year with COVID-19, and the ways the virus has impacted my ability to research and write are incalculable. I know I have been slower, because I, like all of you, have had more work to do—enabling my in-person lessons to work online; running hybrids of online and in-person classes; moving every meeting, defense, and exam onto Zoom; supporting and mentoring students through the pandemic-elevated tension and anxiety—the list of modifications is lengthy, and they all take extra time and labor. I have not been idle during COVID though. In this time, I have written two coauthored academic articles and a book chapter, I have written an article of my own, and I have coedited a special issue of *Enculturation* with my 'Bama colleague Amber Buck. The past two years have not been entirely unproductive, but they have been fueled by anxiety and grief, and the writing and editing of these projects embodies this anxiety and grief. I turned my grief into the production of this book, and I made something akin to a time capsule. I imagine one day I will look back on this work written during this time and wonder how we all managed to function. How we all managed to stay focused and moving forward. We did so imperfectly, of course, but we were there for our students and ourselves. I would not have made it through without my connections on social media—these connections and platforms allowed me to see how others were getting through, sharing suggestions and strategies, and carrying on. I am grateful for all those connections, and my writing on social media connections and identities here carries some of my complex and positive feelings from this pandemic experience. I did not begin this project with a sense of neutrality on social media, and I certainly did not end with one. During this tumultuous time, I relied on social media to combat isolation and anxiety, and at the same time, social media platforms like Facebook were criticized for their toxic moderation practices that fueled conflict and anxiety. Misinformation and disinformation on social media platforms have been a persistent problem, and one that is not easily solved. My relationship status with social media now: it's complicated.

I started writing this book four years ago, and so much has changed since I began. I have grown, and digital Indigenous cultures have grown. The body of Indigenous scholarship in the fields of rhetoric and composition is also growing. The decolonization of myself has been and still is an ongoing project, one made more poignant and personal after the loss of my mother, Helen, in the spring of 2021. She was my teacher and elder, and she left me records and ephemera of the history of our family that have become mine to maintain. She left me with her hopes that I would continue this work of exploring histories and writing about Choctaw women and Indigenous women and continuing the work of Indigenous feminists. Today, I listen to Indigenous podcasts from the US and Canada every day. There are at least three new television shows that feature and revolve around Indigenous stories that I watch as often as I can. I follow Native TikTok. I am a member of even more Indigenous Facebook groups and a follower of more pages than when I started this project. I read Indigenous news daily, and although this is not new, the number of separate publications I read has grown.

As I decolonize myself, I find that the layers of resistance and colonial marginalization fall away from my vision, if not from the positionalities claimed by the institutions in which I live and work. I have said this elsewhere, but I do not like the term *marginalization*, because I think it suggests that the present boundaries can be expanded to bring folx in from the margins. I disagree that stretching the existing paradigm is the answer. We need a new paradigm. As Andrea Riley Mukavetz and I write in our article in volume 9, issue 2, of *Present Tense*, titled "'If You Don't Want Us There, You Don't Get Us': A Statement on Indigenous Visibility and Reconciliation," we need to shatter the old and build a new teaching lodge. And I am building mine with the help of access to digital media and resources. As are, I think and hope, a lot of Indigenous folx, since the number of people who self-identify as Indigenous has almost doubled since the 2010 census.[1] Some of this increase is for certain due to on-the-ground efforts by Native activists to see to it that more Indigenous folx are identified and included in federal censuses, while I expect the rest is the result of access to communities and resources through digital means. Perhaps more people are embracing ancestral Indigenous histories and identities because the country is moving toward embracing multiculturalism. Which, I think, is one of the reasons why this book matters. Because it marks some instances of indigeneity in digital spaces and networks, and it discusses how and why they reflect contemporary notions of indigeneity in digital public discourse, discourse that is expanding to include more people.

In 2021, stories of the "discovery" of Indigenous graves at residential and boarding schools across Turtle Island made international news. It is with a heavy heart that I have watched as the violence and trauma of Canadian residential schools and US boarding schools has circled back through public awareness as unmarked graves are being located with the assistance of activists, archivists, and ground-penetrating technologies. The new media uses language like "discovered," as in "we have discovered these unmarked graves," but of course, Indigenous folks have always known that our children, those who were stolen from us and never heard from again, were buried on those sites. My grandmother told a few stories of her time at school. And while some of her experiences were positive, and Choctaw schools are generally seen as positive, not all of her stories were the good kind. The awareness that the Indian schools were settler-colonial projects of Indian erasure has been part of my family history, like so many other Indigenous families. We know. We have always known.

Also in the news has been the story that the bones of the ancestors, in addition to being buried on school grounds, are also being held by public institutions for "research" and historic purposes. Some remains of these ancestors have already been repatriated, like some of those found at Carlisle Industrial School, while my former institution, the University of Alabama, continues to stall on the repatriation of hundreds of Natives' remains in their collections. The project is enormous, and perhaps a well-earned reckoning of sorts has arrived. I do not know. My research assistant, Emily Wieman, a master's student in our University of Alabama Composition, Rhetorics, and English Studies (CRES) program, and I are collecting data on the public discourse around these (not) "new" discoveries. We are in the early stages of this data collection, but we have uncovered much more settler language and many more settler strategies than I discuss in this manuscript. As I think about continuing this work, I want to reach for ways to make what we find more accessible via networked technologies. I am thinking about how, as a Choctaw woman, I can join my work to the digital activism and digital community of Indigenous folx I discuss in this book and those I have discovered since.

One of the limitations of sharing Indigenous resources and cultures online is that many reservations and Indian communities do not have access to broadband technologies. The Biden administration's infrastructure plan, 2021's American Rescue Plan, sometimes called Build Back Better, promises money to improve broadband access across Indian Country. I will watch and wait as this legislative package stalls in Congress. I hope to see the near future bring this money and expansion. I do not discuss accessibility or the lack of resources for Indigenous digital life, which is especially significant to Indians living on reservations. This lack of access is a significant detriment to engaging in digital culture, and while the reservation system is widely documented as rooted in the federal plan for Indigenous genocide, this lack of investment in Native digital spaces is not only due to the Indian Removal Act but is also bound up in the broader capitalism of the US imagination. If you live in remote spaces where it is not profitable to build broadband, or if you simply cannot afford to pay for high-speed internet yourself, you deserve to go without. This imagination places the blame for poverty and discrimination squarely on the shoulders of those who are most impacted by it. This thorny subject of accessibility is beyond the scope of this book, as I have focused on those who have access to digital culture. I, like many other Indigenous folx, will watch the Build Back Better plan as it continues to develop and change. I hope to come back to this topic later with better news than the situation we have now.

I shift here to discussing what I think are two important things in the field of Indigenous rhetorics that have happened since I began this book. For myself, the more I decolonize my thinking and being, the more I am able to read Indigenous peoples and their discourses and histories and hear Indigenous voices as they share their own lifeways. Through careful reading and study, both the texts I examine here and the scholarship I ground it in, I am improving in my ability to prioritize the voices of Native folx through rhetorical reading and listening. This book project had an earlier title: "Listen." And several of the talks I have given prior to the publication of this book, in which have shared early research with audiences for feedback and kin collaboration, had "Listen" in their titles. I chose *listen* as a watchword because audiences listened to me and I listened to them, and we shared our thoughts on what I have found. And although this is a single-author project, Indigenous knowledges are not made by an individual but emerge from, as I have discussed in this book, collaborative knowledge-making practices—a lot of speaking and even more listening. So, the first thing I think is important in Indigenous rhetorics is that the collaborative listening and thinking in community that my Native cousins in scholarship have been writing about are gaining a kind of critical mass as the field of rhetorics makes its turn toward social justice and cultural situatedness. These turns in the field have opened space for what in the past have been, in many ways, the marginal discourses, methodologies, approaches, pedagogies, and scholarly practices of Indigenous scholars. Especially with the growing interest in cultural rhetorics, we are no longer on those margins.

So much has happened, and so much has changed, but the effectiveness of listening to others and decentering whiteness in the practice of decolonization has not. And this brings me to the second of what I think are major changes in support of Indigenous rhetorics and scholarship in our field. I think an important shift is that there is more purposeful Indigenous hiring and more Indigenous academic programming in our scholarly institutions, particularly in Canada but also increasingly in the United States, as I have observed several Indigenous "cluster hires" in US and Canadian institutions to build cohorts of Native scholars. This hiring and support represents perhaps a kind of institutional commitment to diversity and inclusivity (DEI initiatives) that tends to reinforce the marginal model I mentioned above. However, these changes also create cracks allowing some of us to be able to bring decolonialism into our institutions in limited ways. These decolonization efforts hopefully create more cracks in institutional settler colonialism as we work to *unsettle* it. Of course, this is not true for all of us. There is still plenty of institutional resistance, as Indigenous academics are hired without the support structure, understanding, and redefinition of what work looks like to include Indigenous research and teaching. I think about my own experiences and contemplate a future collaborative project on what makes a successful Indigenous academic in a productive institution. In the meantime, I think my Indigenous colleagues and I will have to make our own spaces and support each other. I am grateful to the Conference on College Composition and Communication Native American Special Interest Group (SIG) for creating a space for us, and I am grateful for everyone who said I could write this book and find an audience for it.

I confess that the theoretical framing in chapter 1 of this book was the most difficult part of the writing for me. I have heard it said that Indigenous folx do not theorize, but we do, as Brayboy and so many others I have cited remind us. Obviously, I discuss storytelling at length, for example. And I weave stories throughout this work. Here is a very short story: I remember in graduate school having several professors explain to me that the knowledge we create might not have an immediate application, but rather might be useful to the next person who might build upon it. And that knowledge creation is a lot like a donation or a deposit into a shared account. Indigenous folx do that too, but with co-created knowledge, it is more like a communal campfire. We talk together, listen with each other, and construct knowledge between people. Or, say, we do not need fire metaphors, and we agree that there are more than a few differences between Indigenous and western approaches to knowledge-making. In chapter 1, I tried to navigate theorizing an Indigenous framework, as one scholar in a context that requires the voices of more than one person for aims other than making new knowledge. Indigenous knowledge-making carries on its back traditions, cultures, and lifeways that reinvest the theorizing into practice. In other words, I cannot bank knowledge without a connection to a lifeway of supporting, sustaining, and protecting the community. For example, Native feminists and Indigenous activists are trying to save everyone. By protecting water and defending land, by demanding justice for children in unmarked graves and missing women who have never been found, Native feminists have their eyes set on sharing rather than banking. There is a lot of work to do. So, in chapter 1, I try to wrangle together a set of theories that will undermine settler-colonial interpretations of Indigenous rhetorics and identity constructions, to bring to the forefront Indigenous voices. Because, again, we are trying to save everyone. If I am not entirely successful, I at least hope to add to conversations with other Indigenous scholars around theorizing the field.

Beyond Research Questions

I began this project with two research questions:

- How do Indigenous peoples construct themselves in digital spaces and places, as opposed to how digital medias construct them?
- Can I find other examples of where the stereotypical descriptions of the race of Indigenous peoples are complicated or subsumed in digital spaces?

Beginning with chapter 1, I theorized the intersections of white supremacy, settler colonialism, neoliberalism that informs the design of digital technologies, and digital rhetorics, which set up the framework for the case studies. I think in this book I have done more than answer these two questions. I have provided examples of online Indigenous identity construction, and I have shared the ways these identities create digital communities. I have also described some of the reasons Indigenous folx might create these communities, both for cultural motivations and socially just motivations. I have discussed the problems of white supremacy and settler colonialism in the commonplace interpretations of Indigenous peoples, their lives, and their relevance in modern life.

In my chapter 2 discussion of the #MeToo movement, I used Indigenous identity construction to create the space for my conversation about #MeToo as a digital social movement, and I offered an Indigenous way of knowing as one approach to thinking about the discussions of and reactions to #MeToo that nuanced these discussions beyond binaries of bad and good and privileged and marginalized, and beyond white supremacist interpretations. I argued, speaking with the power of a Choctaw woman rather than accept patriarchal marginalization, that listening with a good heart and in good intent ruptures the patriarchal paradigm that women cannot be trusted to be truthful and that women bring the unwanted sexual advances of men onto themselves by ther nature. Here, listening is taken beyond the patriarchal power dynamic as the social media posts become stories and personal stories serve as evidence, not in the true/false paradigm of western logic but in the veracity of experience and the learning that is situated within it.

Also in chapter 2, I laid out my methodology that I would use to approach my case studies from Indigenous lifeways and knowledge-making practice. I wrote chapter 2 while I still had my mother, and my lineage influences this chapter, because I situate myself as a woman in a line of women who have agency to speak to white supremacy and settler colonialism. I mention the loss of my mother many times in this work, and I feel that loss daily. There are so many more things I wish I could ask her and so many of her stories that went untold. I regret not asking her to write more of her own work, because she was a writer too and also a voracious reader. I owe my own commitment to these things to her.

Chapter 3's examinations of Indigenous political and identity memes and their distribution and reception on a member-only Facebook page expanded my scope on how Indigenous people encounter white supremacy and settler colonialism as systems of oppression. There are memes that celebrate the power of women, that condemn the militarization of the police, and that call out the infection of settler colonialism in our own Indigenous communities. The memes in chapter 3 themselves relate to my second research question in that they are created and circulated by Indigenous people for Indigenous people, and their existence complicates and nuances white perceptions of Native peoples.

In chapter 4, I speak with my friend Jeffrey Veregge about his digital artistry and his social media presence. This chapter is a look at the work and perspectives of an Indigenous artist, but it is also an example of two Indigenous people making relations and making meaning together. As I write this, Veregge continues his battle with lupus and is being evaluated for a potential liver and kidney transplant. His illness is very serious, and I deeply appreciate that I was able to talk to him and make community with him.

In chapter 5, I look back carefully at the white supremacy and settler-colonial capitalism around Payu Harris's Bitcoin project. Again, while this was the first attempt at an Indigenous cryptocurrency, it likely will not be the last. In a recent discussion I had with a potential collaborator, I answered her question as to whether I thought there could be a positive outcome where capitalism intersects with Indigenous lifeways, particularly in the digital economy. I actually do think this is possible in some carefully considered ways, because I witness it weekly as I tune into Indigenous jewelry shows featuring community artists and live streamed on Facebook. I have purchased jewelry for myself through these community showings. The potential for growth and reach in Indigenous people's economies through live streaming, on online shopping platforms, and for the social networking of Indigenous arts and artists inspires me. Payu Harris inspires me.

Implications for Future Research

The methods, theories, and frameworks described here are very much the beginning of my work in thinking about decolonizing digital identity and meaning-making practices. There is a lot more here to do, particularly in terms of Indigenous activism, but around Indigenous representation and knowledge-making as well.

I think decolonizing digital identity work is particularly useful in Indigenous activism, because it could lead to greater understanding between parties and greater impacts of this labor. Here I offer the example of the hashtag #waterislife (meaning "water is life") as it relates to Indigenous activism against oil pipelines that cross sacred lands and waterways. I am speaking generally here of activists and their allies, who have many water-related aims and goals, but I write here on the subject of Water Protectors who tag their protest and advocacy work with #waterislife. I take this moment to unpack "water is life." First, what response does a white audience have to the phrase *water is life*? I cannot ask everyone, so I asked my undergraduates this semester what *water is life* meant to them, and they replied from a western perspective that water is a significant portion of our earth and our atmosphere, that our bodies are largely made up of water, and that we all need water to live. They all agreed that water is incredibly important to life and were concerned with water quality, and in a few cases, access to clean water. They are an informed group of people, and they have all expressed concern for climate change and the destabilization of our environment from things like clear-cutting and industrial pollution. And I agree that *water is life* does mean these things. But to an Indigenous Water Protector, it also means that water is life, literally that the water is alive and has kinship with the people who rely on it and protect it. To different Indigenous peoples, that life that water has can be encountered in different ways, but that the water is alive, that its life is sacred, and that we have a kinship responsibility to protect and sustain it as it reciprocally protects and sustains us is a commonly held idea. The idea of what sacredness is and what *sacred* means is also different—perhaps subtly, but different—for Indigenous peoples. Western notions of the sacred and sacrality are generally religious in nature but can also be civic and cultural. Thus, beliefs that are sacrosanct, like individual liberty is to Americans as laid out in foundational documents. For Indigenous people, animals, land, and water are all sacred. They are not unapproachable, like those religious objects or civil principles, they are respected and valued members of the community. We are related to them, and through that relationship we can protect them.

Circling back to pipeline protestors: water is important, but to truly understand the aims of the activists, you must know that water is life. That life is sacred. And that polluting that water threatens its life and violates its sacredness. This is before we consider how polluting the water impacts fragile ecosystems and the people who need clean water to drink. To understand why this distinction matters in protests, I return to my example of the understanding of my socially conscious undergraduates, that to them water is crucial to human life. However, in their paradigm, many things are critical to human life, like oil and gasoline, electricity, cars and planes, and military defense applications. Balancing these critical needs is a perspective from the settler-colonial model, because settlers possess all that they have conquered, and it is within their purview to sacrifice some assets in favor of others. Indigenous folx are saying instead that a violation of the water parallels the violation of a person. That water is not a resource only; rather, it is a relation and relations are sacred. I argue here that if white audiences understood the lifeways of Indigenous peoples, they could gain a different perspective of water outside the settler-colonial paradigm, which might just lead to new opportunities for conservation. Therefore, I think it is important to decolonize public thinking when it comes to activism, to understand the messages of the activists from their own perspectives rather than a settler-colonial one that perpetuates the consumption of resources. My future projects work in decolonial directions.

This project has also helped me to think more about how Indigenous folx use digital technologies to create and sustain their communities. I have in 2021 and 2022 listened to many Indigenous podcasts produced across Turtle Island. The topics of these podcasts extend from Indigenous news and culture, current events across Indian Country, Indigenous causes, humor, and awareness-raising. I think podcasting is a rich area for research, and I would like to interview podcasters for their take on the community and identity work, particularly in the activism I have mentioned, including pipeline activism, Missing and Murdered Indigenous Women (MMIW), the discourses of and around repatriation of Indigenous remains (NAGPRA, the Native American Graves Protection and Repatriation Act), and Canada's Truth and Reconciliation project. It has also given me the time and space to think about Indigenous environmentalism and the use of social media to promote and educate about these lifeways.

I regularly use the Choctaw Nation of Oklahoma's website for news and information and for cultural programming. I use the Nation's YouTube channel, which features council meetings, storytellers, and language classes. I have enrolled in a language class that is taught online, and I hope to enroll in more. The website also has a member portal that allows members to access tribal programs and services. Tribal websites, like most government websites, are important sources of information. I have accessed several as I worked on this book, as they provide access to historical information not easily available otherwise. I look forward to the decolonial work that can be done as I investigate these and other topics.

Research Contributions

I wrote this book as a narrative of my development as an Indigenous scholar, along with the development of methods and frames for Indigenous research. It is a personal manuscript, one that is informed by intersectional feminism, Indigenous feminism, and Indigenous lifeways and worldviews. I mean this to say that the work is personal, because Indigenous work is personal, feminist work is personal, and the work of social justice and equity are personal. This is a work of storytelling, and stories are always personal. As I wrote this story, I also wrote the story of my family, which was not one of my original goals. This happened in large part because losing my mother was a transformational event in my life, in that I must learn to live without her and I must assume some family responsibilities in her place. I see my motherless self differently now by writing this story. I see my Choctaw self differently now too—I am more certain of my voice, my purpose, and my place in my communities.

In terms of what this work contributes to the scholarly conversations around digital identity, digital platforms, circulation and virality, networking in multiple meanings, internet studies, digital rhetoric, and writing studies, I contend that I add nuance to the raced white supremacy of these conversations. While it might be obvious that white supremacy has attempted to and continues to attempt to erase Indigenous peoples and eradicate our cultures and histories, I think my work on memes, for example, shines light on the way white supremacy inflicts its perspective and how it is possible to undermine and complicate that perspective using those memes as identity-building, community-supporting tools. It is not the goal of creating the meme to send it out and see it attain maximum circulation; rather, the memes express common experiences for specific audiences to build the identity and community of that audience. Through this example and others, I demonstrate that Indigenous folx are not relegated to the past, but we are actively working with and using digitally mediated communication and meaning-making strategies to resist erasure and the commonplace interpretation of Indigenous peoples as relegated to the past rather than as digital makers of the present.

I also think this work adds to the growing body of scholarship in Indigenous methods and approaches to research. While there are several books on Indigenous methodology, they have not been adapted to think about and analyze digital spaces. So, I think this book offers methodological tools and practices to the field in at least three different ways. The first way is that I am offering some theory and practice to demonstrate Indigenous ways of being and knowing that can be applied to decolonize the white supremacist and settler-colonial gaze, offering a broader perspective in understanding indigeneity today. The second is specific to digital media and artifacts produced by and for Indigenous people—in that the approaches I describe here can be used to better understand this way of being in the digital world. Finally, I propose that these decolonial practices can be more broadly applied, like I do in chapter 1, to other studies of digital community and identity practices beyond Indigenous peoples. I opened this book by conveying the knowledge that only 2 percent of the US population is Indigenous, although that number is growing thanks in part to on-the-ground census work in Native communities as well as a growing interest and acceptance of multiculturalism. I suggested that studying 2 percent of the population might seem like a small sample of a marginalized people, but I argued that these practices and lifeways that I am studying can be more broadly applied. Here I contend that breaking outside the white supremacist and settler-colonial models can inform approaches to addressing a number of social issues, including climate change, violence against women, and environmental justice.

I want to end on research and pedagogical suggestions because I learned many things while writing this book about both. The first thing I learned, and the thing I struggled the most with, is writing a monograph itself. A monograph is a single-authored project, in my case required for tenure, and I at first did not like working on it alone. Indigenous rhetorics, as I have repeatedly written, are made in community, so writing a single-authored book about them is strange and uncomfortable. The case studies are collective projects—social media activism, memes, interviews, and mass media analysis that began as a collaborative project. I wrote a monograph because it is expected, but I wrote it about many people doing digital rhetoric in different ways. I struggled repeatedly with isolation and a sense of disconnectedness from my communities. Things got easier for me when I began meeting with other people to write. I wrote with colleagues and graduate students as well as online writing and accountability groups. In hindsight, the struggle to get this book out and my feelings of isolation make sense, but during those early months, I struggled. I also found that taking on collaborative projects while writing the monograph helped. A few of those projects were Indigenous and others were not, but together they filled my need to think with other people. Writing in collaboration gave me opportunities to talk through ideas and share space and support. All the Indigenous researchers I have read and have worked with emphasize good and right relations and their centrality to Native identity and knowledge practices. There is a reason for that. So, the first thing is not to write a monograph alone—write it in concert with other writers, especially if you are Native and doing Indigenous work. We are not meant to be thinking things through only by and for ourselves.

Another important thing I want to note in Indigenous research methods is accountability. If being in relation to the land, the natural world, and other humans is central to Native thinking (Wilson 2008, 73–74), then accountability to those relationships is critical to Indigenous praxis. This is more than a procedure in that we think about western research as repeatable and verifiable. It is a worldview: "Rather than the goals of validity and reliability, research from an Indigenous paradigm should aim to be authentic or credible . . .

the research must accurately reflect and build up on the relationships between the ideas and the participants. The analysis must be true to the voices of all participants and reflect an understanding of the topic that is shared by researcher and participants alike" (Kovach 2009, 101–2). In other words, other than relying on "fact" as it is defined in the verifiability-testability paradigm of "science," Indigenous folx rely on the authenticity of ties between participants and researchers. It is about trust; therefore, to have trust, you have to be accountable to each other and to your community. In an upcoming edited collection, I write about the importance of relationships and accountability in my mentoring practices, but I think accountability is even more important in research methodology because we center people, not process. The research and writing process does not reinforce isolation but community. This too can help address the isolation that often characterizes graduate school and the tenure process.

There are a number of excellent books on Indigenous research methods that I have cited in this work. The most important thing about these methods, I think, is described by Kovach: "In conceptualizing a tribal methodology, I have identified a theoretical positioning as having its basis in critical theory with a decolonizing aim in that there is a commitment to praxis and social justice for Indigenous people" (2021, 48). In other words, we do Indigenous research and apply Indigenous methods because we prioritize social justice. In working through this book, I thought a lot about social justice and the outcomes I would like to see in the future. I want to see ruptures in patriarchy that allow women to claim their power in speaking into the public sphere. I want to see digitally literate practices continue to increase among Indigenous peoples, including improved access to the digital tools and platforms. I want to see a broader public awareness of Indigenous issues like MMIW, land back, and others. I want to see Indigenous knowledges and practices taken up in public policy discussions, like forest management and water conservation. I want to protect voting rights and improve political representation. I want physical things too, like for elder remains to be repatriated and for institutions to stop cutting the hair of Indigenous peoples without their consent. I want to see Indigenous histories and knowledges taught in schools, not just in designated months or special topics courses, but I want to decolonize education so that settler colonialism is not the default worldview taught in schools.

As I finish writing this manuscript, I want to start it over again. There are things I would do differently, because I have grown and changed in this process, and I know a lot more about what I want for myself and my work. I would change the stance of this book so that I focused on a Native audience rather than a more general scholarly audience. I think that would demonstrate my confidence in the relevance and significance of these ideas. I do not want to see my writing as an effort to persuade readers that these methods and practices matter. Because they do matter. Also, I would cite this book more deeply. As I have written this, I have read more and more, and since I have completed writing various chapters, I continue to read. If I keep revising it for readings I have read, talks I have heard, and conversations I have participated in, I might never publish it. But the impulse is there. In the third edition of *Decolonizing Methodologies*, Smith writes about Indigenous students feeling constrained to put knowledge on the page, not just because of respect and reciprocity but also because of identity and the negotiations of place in the community (154–55). I felt that as I began the project, and I suspect some measure of this remains in wanting to cite more and more researchers and teachers.

I hope you have learned something reading it. I hope you take away that relationships are central to worldview, research stories are knowledge-making, and knowledge-making is done in accountable community. I hope you see the richness of Indigenous digital practices and broaden your experience with Indigenous issues and ideas. I hope you know I am grateful for your attention and your time.

Notes

Introduction: Doing Storytelling as Epistemology

1. Choctaw is the English name of our tribe, the name the federal government uses. Chahta is our name in our own language. We are also known historically as the Red People and Forest People, and as those who have flat foreheads, according to various histories I have read and stories I have been told. I have tried to be consistent in my use of Chahta and Choctaw, using our own language when I am speaking personally and the English name when referencing texts like federal sources and resources from the Nation ourselves. In short, if I am talking about the culture, it is Chahta; where I am referring to a federally recognized body and its institutions, it is Choctaw. Where I cite treaties, I represent the naming in the original texts.
2. "Member" or "citizen"; although, some prefer *citizen* as a reflection of sovereignty.
3. That North America grew on the back of a giant turtle, called Turtle Island, is a common oral history among people indigenous to North America.
4. My father is English, Dutch, and Cherokee, among others. He introduced me to science fiction and classic monster movies, nurtured my creative spirit, and gave me my first computer. Although I do not talk about him in this project, he is more than a mere footnote in my life.
5. *Indian*, *Indianness*, *NDN*, and other variations are sometimes used by Native American peoples in the United States to describe themselves. These terms are often inherited from family members, and to use them is to use the language of one's own people. I use these terms.
6. The Dawes Rolls or "Final Rolls" are a census (or inventory, really), created from 1898 to 1914 to purportedly identify all the Native American people from the Cherokees, Choctaws, Chickasaws, Creeks, and Seminoles for federal "recognition." I include this note as

informational only, and I strongly reject the idea that a federal inventory should be the final authority on who is Native American and who is not.
7. Blood quantum is extremely problematic, and using it in the way I am describing it, as a racial identification construct, is racist. Here I reiterate the trap of Indigenous racial identity, where blood quantum is used both to identify tribal membership and to inflict settler violence on Indigenous peoples. This process has led to the exclusion of many people, including Black people who were enslaved by some Native tribes before emancipation.
8. This claim was much researched at the time, and that research is beyond the scope of this book. However, it is important to note that President Trump's many claims about Warren using her Cherokee family anecdotes to her advantage are false.
9. In my own family, one of my great-grandfathers was not included in the rolls. We know he was Choctaw, like his wife, my grandmother, my mother, and I, all with Choctaw enrollments. But the reason for his absence from the records is a long-lost story in our family history. This absence is not uncommon given the wide array of reasons Indigenous people may not have been present or may not have been determined to be Indigenous enough by the Dawes Commission, who made these determinations. These nuances are not to say I support Warren's claims, because I do not, but that I understand identity is more complex and nuanced than campaign talking points.
10. At the time of this writing, Trump has made references to Warren as "Pocahontas" and made dozens of disparaging remarks about her false indigeneity. Pocahontas was the daughter of chief Powhatan, a leader of an extended network of tribal peoples in and around the area of Jamestown, the very early, seventeenth-century community of English settlers in North America. According to some sources, Pocahontas was seventeen years old when she was kidnapped by settlers in a period of ongoing conflict with her father Powhatan's peoples. Her story has been subsumed into the settler-colonial myths of nation-making (Townsend 2004).

Chapter 1: Indigenous Storytelling and Ways of Thinking and Being

1. "Tanchi" is the spelling for corn used in the Choctaw Dictionary (https://dictionary.choctawnation.com/word/) and other stories I have found. I am unsure if "tunchi" is an alternate spelling or an error introduced during transcription.

Chapter 2: Listen

1. Neither a video nor audio recording of this talk is currently available. This pre-pandemic conference did not include the hybrid elements we have come to expect from post-pandemic academic conferences and talks.
2. On May 9, 2023, a jury found Donald Trump liable for sexually abusing E. Jean Carroll. The jury awarded Carroll $5 million in damages for the assault and for damaging comments Trump made about her in the media (Neumeister, Peltz, and Sisak 2023).
3. I recall once taking two days off from my job to attend the funeral events of my cousin. My employer was surprised (and frustrated), saying that we must be a "weirdly" close family. I still do not think it is odd to make an effort to attend to cousins as valued members of the family, even if I did see them perhaps more often than traditional American nuclear families visit their own relatives.

Chapter 4: Jeffrey Veregge

1. Author interview with Jeffrey Veregge on September 15, 2020.
2. Here, I choose to address a new relation by calling him by his first name instead of using his last name, which is more traditional in academic practice. I chose one tradition over another.
3. Chief Wahoo was a redface caricature of a Native American man, used as a mascot from the 1950s to 2018 by the Cleveland Indians, a professional baseball team in Cleveland, Ohio. Chief Wahoo was abandoned in 2018 after a public movement of protest raising awareness of racist mascots for sports teams and schools.
4. At the time of this writing, in September 2020, there are hundreds of wildfires raging in the American West.
5. He is referring to Donald Trump here, as this interview was recorded in 2020.
6. Red Wolf is a Marvel superhero of Native American origin. See Leask 2015.

Chapter 5: MazaCoin

This chapter is an elaboration and further analysis of the indigenous cryptocurrency, MazaCoin, that I originally described in an article, "Indigenous cryptocurrency: Affective capitalism and rhetorics of sovereignty," cowritten with my friend and frequent collaborator, John Carter McKnight, that was published in First Monday, volume 21, number 10, on October 3, 2016. While in 2016, John and I were writing from the perspective of how MazaCoin was situated within the global range of local currencies, in this chapter I am concerned specifically with MazaCoin as a resistance and sovereignty movement, its reception as a model of modern settler-colonialism and neoliberalism, and its implications for decolonial research in digital spaces. I write the continuation of this research with John's approval and, here, I thank him for the role he has played in my interest in digital decolonizing narratives and practices.

1. CoinDesk.com is a news and information website that caters specifically to the cryptocurrency community.
2. "Oyate" is an Ojibwa term meaning "the people."
3. According to an online obituary, Dana Lone Hill walked on November 15, 2019, at just forty-seven years old. Pine Ridge, where she lived, has the lowest life expectancy in the United States. She was a writer and an activist.
4. *Russia Today* is now known as "*RT*," and as of 2017, *RT America* must register as a "foreign agent" so that its financial records are transparent. This is because *RT* can serve as a propaganda source for the Russian government and is known to have shared conspiracy theories, fake news, and stories designed to sow discord in the West. It is also a widely seen global news network that provides global stories with a strong pro-Russian bias.

Conclusion

1. From United States Census Bureau, "Selected Population Profiles 2020."

References

Abad-Santos, Alex. 2018. "Nike's Colin Kaepernick Ad Sparked a Boycott—and Earned $6 Billion for Nike." *Vox*, September 24, 2018. https://www.vox.com/2018/9/24/17895704/nike-colin-kaepernick-boycott-6-billion.

Adams, Mikaëla M. 2016. *Who Belongs? Race, Resources, and Tribal Citizenship in the Native South*. New York: Oxford University Press.

AFP. 2018. "US Tourist Killed by Arrow-Shooting Indian Tribe." *France 24*, November 21, 2018. https://www.france24.com/en/20181121-us-tourist-killed-arrow-shooting-indian-tribe.

Akers, Donna. 2013. *Culture and Customs of the Choctaw Indians*. Santa Barbara, CA: Greenwood Press.

Armus, Teo. 2019. "November Is Native American Heritage Month. Critics Say Trump Is Subverting It with a New Celebration of the Founding Fathers." *Washington Post*, November 5, 2019. https://www.washingtonpost.com/nation/2019/11/05/trump-changes-november-native-american-heritage-month-honor-founding-fathers/.

Arola, Kristin L. 2017. "Chapter 11: Indigenous Interfaces." In *Social Writing / Social Media: Publics, Presentations, and Pedagogies*, edited by Douglas M. Walls and Stephanie Vie, 209–24. Fort Collins, CO: WAC Clearinghouse.

Arvin, Maile, Eve Tuck, and Angie Morrill. 2013. "Decolonizing Feminism: Challenging Connections between Settler Colonialism and Heteropatriarchy." *Feminist Formations* 25, no. 1 (Spring): 8–34. https://doi.org/10.1353/ff.2013.0006.

Barlow, John Perry. 1996. "A Declaration of the Independence of Cyberspace." *Electronic Frontier Foundation*, February 8, 1996. https://www.eff.org/cyberspace-independence.

Baym, Nancy K. 2010. *Personal Connections in the Digital Age*. Cambridge: Polity Press.

Boaz, David. 2019. "Key Concepts of Libertarianism." *Cato Institute*, April 12, 2019. https://www.cato.org/publications/commentary/key-concepts-libertarianism.

Booker, Brakkton. 2020. "North Dakota and Native American Tribes Settle Voter ID Lawsuits." *NPR*, February 14, 2020. https://www.npr.org/2020/02/14/806083852/north-dakota-and-native-american-tribes-settle-voter-id-lawsuits.

Bradbury, Danny. 2014. "MazaCoin Aims to Be Sovereign Altcoin for Native Americans." *CoinDesk*, February 5, 2014. http://www.coindesk.com/MazaCoin-sovereign-altcoin-native-americans/.

Bratta, Phil, and Malea Powell. 2016. "Entering the Cultural Rhetorics Conversations." *Enculturation: A Journal of Rhetoric, Writing, and Culture*, no. 21 (April). http://www.enculturation.net/entering-the-cultural-rhetorics-conversations.

Brayboy, Bryan McKinley Jones. 2005. "Toward a Tribal Critical Race Theory in Education." *The Urban Review* 37, no. 5 (March): 425–46. https://doi.org/10.1007/s11256-005-0018-y.

Browning, Lynnley. 2014. "Oglala Sioux Hope Bitcoin Alternative, Mazacoin, Will Change Economic Woes." *Newsweek*, August 14, 2014.

Carson, James Taylor. 1999. *Searching for the Bright Path: The Mississippi Choctaws from Prehistory to Removal*. Lincoln: University of Nebraska Press.

"Chim Afvmmi Na Yupka! Happy Birthday." n.d. Choctaw Nation of Oklahoma.

Clayton, Matt. 2020. *Choctaw Mythology: Captivating Myths from the Choctaw and Other Indigenous Peoples from the Southeastern United States*. Tustin, CA: Refora Publications.

Collins, Patricia Hill, and Sirma Bilge. 2016. *Intersectionality*. Cambridge: Polity.

Consunji, Bianca. 2014. "One Man's Lonely Quest to Build 'Bitcoin for Native Americans.'" *Mashable*, September 18, 2014. http://mashable.com/2014/09/18/MazaCoin-bitcoin-native-americans/#s99u5CwLqkqc.

Cottom, Tressie McMillan. 2019. *Thick: And Other Essays*. New York: The New Press.

Deer, Sarah. 2015. *The Beginning and End of Rape*. Minneapolis: University of Minnesota Press.

Deloria, Vine. 1988. *Custer Died for Your Sins: An Indian Manifesto*. Norman: University of Oklahoma Press.

Diamond, Jared M. 1997. *Guns, Germs, and Steel: The Fates of Human Societies*. New York: Norton.

Domonoske, Camila. 2018. "Many Native IDs Won't Be Accepted at North Dakota Polling Places." *NPR*, October 13, 2018. https://www.npr.org/2018/10/13/657125819/many-native-ids-wont-be-accepted-at-north-dakota-polling-places.

Dougherty, Timothy R. 2016. "Knowing (Y)our Story: Practicing Decolonial Rhetorical History." *Enculturation: A Journal of Rhetoric, Writing, and Culture*, no. 21 (April). http://www.enculturation.net/knowing-your-story.

Downes, Stephen. 1999. "Hacking Memes." *First Monday* 10, no. 4 (October). https://doi.org/10.5210/fm.v4i10.694.

Dunbar-Ortiz, Roxanne. 2014. *An Indigenous Peoples' History of the United States*. Boston, MA: Beacon.

Dunphy, Peter. 2019. "The State of Native American Voting Rights." *Brennan Center for Justice*, March 13, 2019. https://www.brennancenter.org/blog/state-native-american-voting-rights.

Ecoffey, Brandon. 2014. "Oglala Sioux Tribe Surprised by MazaCoin Plan." *Native Sun News*, March 7, 2014. www.indianz.com/News/2014/012781.asp.

Edwards, Helen Jane. n.d. Personal archives: "Histories and Stories." Tempe, AZ.

Fonseca, Felicia. 2020. "Warren Still Dogged by Past Claims of Indigenous Ancestry." *Fox 11 News*, February 27, 2020. https://fox11online.com/news/connect-to-congress/warren-still-dogged-by-past-claims-of-indigenous-ancestry.

Friel, Katie, and Emil Mella Pablo. 2022. "How Voter Suppression Laws Target Native Americans." *Brennan Center for Justice*, May 23, 2022. https://www.brennancenter.org/our-work/research-reports/how-voter-suppression-laws-target-native-americans.

Gallway, Alexander R., and Eugene Thacker. 2007. *The Exploit: A Theory of Networks*. Minneapolis: University of Minnesota Press.

Gee, James Paul. 2009. "10—Semiotic Social Spaces and Affinity Spaces: From the Age of Mythology to Today's Schools." In *Beyond Communities of Practice: Language, Power, and Social Context*, edited by David Barton and Karin Tusting, 214–32. Cambridge: Cambridge University Press.

Geertz, Clifford. 1973. *The Interpretation of Cultures: Selected Essays*. New York: Routledge.

Gilyard, Keith, ed. 1999. *Race, Rhetoric, and Composition*. Portsmouth: Heinemann.

Glancy, Diane, and Linda Rodriguez. 2023. *Unpapered: Writers Consider Native American Identity and Cultural Belonging*. Lincoln: University of Nebraska Press.

Greenfield, Adam. 2017. *Radical Technologies: The Design of Everyday Life*. London: Verso.

Gries, Laurie E., and Collin Gifford Brooke. 2018. *Circulation, Writing, and Rhetoric*. Logan: Utah State University Press.

Grim, John A. 2009. "6: Indigenous Lifeways and Knowing the World." *The Oxford Handbook of Science and Religion*, edited by Philip Clayton, 87–107. Oxford: Oxford University Press. https://doi.org/10.1093/oxfordhb/9780199543656.003.0007.

Haas, Angela M. 2007. "Wampum as Hypertext: An American Indian Intellectual Tradition of Multimedia Theory and Practice." *Studies in American Indian Literatures* 19, no. 4, series 2 (Winter): 77–100. http://www.jstor.org/stable/20737390.

Hall, Jacquelyn Dowd. 2005. "The Long Civil Rights Movement and the Political Uses of the Past." *The Journal of American History* 91, no. 4 (March): 1233–63. https://doi.org/10.2307/3660172.

Haltiwanger, John. 2017. "Who Are the Navajo Code Talkers? Trump Insults World War II Heroes with 'Pocahontas' Joke." *Newsweek*, November 27, 2017. https://www.newsweek.com/who-are-navajo-code-talkers-trump-insults-wwii-heroes-pocahontas-joke-723821.

Hamill, Jasper. 2014. "The Battle of Little Bitcoin: Native American Tribe Launches Its Own Cryptocurrency." *Forbes*, February 27, 2014. https://www.forbes.com

/sites/jasperhamill/2014/02/27/the-battle-of-little-bitcoin-native-american-tribe-launches-its-own-cryptocurrency/#49d381a47c5b.

Haraway, Donna J. 2016. *Staying with the Trouble: Making Kin in the Chthulucene*. London: Duke University Press.

Hayes, Elisabeth, and James Paul Gee. 2010. "Chapter 21: Public Pedagogy through Video Games: Design, Resources, and Affinity Spaces." In *Handbook of Public Pedagogy*, edited by Jennifer A Sandlin, Brian D. Schultz, and Jake Burdick, 185–93. Routledge Handbooks Online.

Hawbaker, K. T. 2019. "#MeToo: A Timeline of Events." *Chicago Tribune*. Accessed January 24, 2019. https://www.chicagotribune.com/lifestyles/ct-me-too-timeline-20171208-htmlstory.html.

Hayoun, Massoud. 2018. "For Native Americans, the Battle against Voter Suppression in North Dakota Is Only the Beginning." *Pacific Standard*, November 6, 2018. https://psmag.com/social-justice/for-native-americans-the-battle-against-voter-suppression-in-north-dakota-is-only-the-beginning.

Jensen, Tracy. 2013. "A Summer of Television Poverty Porn." *The Sociological Imagination*, September 9, 2013. https://web.archive.org/web/20161030064805/http://sociologicalimagination.org:80/archives/14013/comment-page-1.

Jensen, Tracy. 2014. "Welfare Commonsense, Poverty Porn and Doxosophy." *Sociological Research Online* 19 (3): 277–83. https://doi.org/10.5153/sro.3441.

Kessler, Glenn. 2018. "Just About Everything You've Read on the Warren DNA Test Is Wrong." *Washington Post*, October 18, 2018. https://www.washingtonpost.com/politics/2018/10/18/just-about-everything-youve-read-warren-dna-test-is-wrong/.

King, Lisa, Rose Gubele, and Joyce Rain Anderson, eds. 2015. *Survivance, Sovereignty, and Story*. Logan: Utah State University Press.

King, Thomas. 2008. *The Truth about Stories: A Native Narrative*. Minneapolis: University of Minnesota Press.

Knobel, Michele, and Colin Lankshear. 2007. "Online Memes, Affinities, and Cultural Production." In *A New Literacies Sampler*, edited by Michele Knobel and Colin Lankshear, 199–227. New York: Peter Lang.

Kovach, Margaret. 2009. *Indigenous Methodologies: Characteristics, Conversations, and Contexts*. 1st ed. Toronto: University of Toronto Press.

Kranzberg, Melvin. 1986. "Technology and History: 'Kranzberg's Laws.'" *Technology and Culture* 27, no. 3 (July): 544–60. https://doi.org/10.2307/3105385.

"Lakota Nation Adopts MazaCoin Crypto-Currency as Legal Tender." 2014. *RT News*, March 3, 2014. https://www.rt.com/usa/native-american-nation-bitcoin-632/.

Landry, Alysa. 2014. "Questions Surrounding MazaCoin, the Lakota Cryptocurrency: Answered." *Indian Country Today*. March 3, 2014. https://ictnews.org/archive/9-questions-surrounding-mazacoin-the-lakota-cryptocurrency-answered.

Leask, James. 2015. "Not So New, Not So Different: On *Red Wolf* and Indigenous Representation in the New Marvel." *ComiczAlliance*, June 5, 2015. https://comicsalliance.com/all-new-all-different-red-wolf-marvel/.

Leavy, Patricia, and Anne Harris. 2019. *Contemporary Feminist Research from Theory to Practice*. New York: Guilford Press.

Lee, Michelle Ye Hee. 2016. "Why Donald Trump Calls Elizabeth Warren 'Pocahontas.'" *Washington Post*, June 28, 2016. https://www.washingtonpost.com/news/fact-checker/wp/2016/06/28/why-donald-trump-calls-elizabeth-warren-pocahontas/.

Leonard, Michele. 2023. "You Don't Look Indian." *Unpapered: Writers Consider Native American Identity and Cultural Belonging*, edited by Diane Glancy and Linda Rodriguez, 125–36. Lincoln: University of Nebraska Press.

Markham, Annette N. 2004. "Internet Communication as a Tool for Qualitative Research." In *Qualitative Research: Theory, Method and Practice*, edited by David Silverman, 95–124. London: Sage.

Marvel. 2020. "The Demon Rider Returns on Jeffrey Veregge's Spellbinding 'Marvel's Voices: Indigenous Voices' #1 Cover." October 21, 2020. https://www.marvel.com/articles/comics/the-demon-rider-returns-on-jeffrey-veregge-s-spellbinding-marvel-s-voices-indigenous-voices-1-cover.

Marwick, Alice E. 2015. *Status Update: Celebrity, Publicity, and Branding in the Social Media Age*. New Haven, CT: Yale University Press.

"Maza." 2016. Maza-Online. https://web.archive.org/web/20160804051820/http://www.maza-online.com/index.html.

McKee, Heidi, and James Porter. 2009. *The Ethics of Internet Research: A Rhetorical, Case-Based Process*. New York: Peter Lang.

Merrill, Dave, and Lauren Leatherby. 2018. "Here's How America Uses Its Land." *Bloomberg*, July 31, 2018. https://www.bloomberg.com/graphics/2018-us-land-use/#xj4y7vzkg.

"#MeToo: A Timeline of Events." 2017. *Chicago Tribune*, December 8, 2017. https://www.chicagotribune.com/lifestyles/ct-me-too-timeline-20171208-htmlstory.html.

Milano, Alyssa (@Alyssa_Milano). 2017. "If You've Been Sexually Harassed or Assaulted Write 'Me Too' as a Reply to This Tweet." X (Twitter), October 15, 2017. https://twitter.com/alyssa_milano/status/919659438700670976?lang=en.

Nahdee, Ali. 2020a. "The 'Aila Test' Evaluates Representation of Indigenous Women in Media." Interview by Shea Vassar. *High Country News*, May 14, 2020. https://www.hcn.org/articles/indigenous-affairs-interview-the-aila-test-evaluates-representation-of-indigenous-women-in-media?fbclid=IwAR1R83qmFVVY3IeM-tr6qclgeMy1uoR-xB-H0HO26YKtNiA7VF0H50d76Ac.

Nahdee, Ali (@the-aila-test). 2020b. "Why Disney Should Remove (and Apologize for) *Pocahontas*." Tumblr, June 26, 2020. https://the-aila-test.tumblr.com/post/622042408078704640/why-disney-should-cancel-and-apologize-for.

Nelson, Alondra. 2018. "Elizabeth Warren and the Folly of Genetic Ancestry Tests." *New York Times*, October 17, 2018. https://www.nytimes.com/2018/10/17/opinion/elizabeth-warren-and-the-folly-of-genetic-ancestry-tests.html.

Neumeister, Larry, Jennifer Peltz, and Michael R. Sisak. 2023. "Jury Finds Trump Liable for Sexual Abuse, Awards Accuser $5M." *Associated Press*, May 9, 2023. https://apnews.com/article/trump-rape-carroll-trial-fe68259a4b98bb3947d42af9ec83d7db.

Newitz, Annalee. 2013. "Epic Superhero Art in a Traditional Native American Style." *Gizmodo*, Apr 11, 2013. https://io9.gizmodo.com/amazing-superhero-art-in-the-style-of-native-american-472345304.

Niezen, Ronald. 2005. *Digital Identity: The Construction of Virtual Selfhood in the Indigenous Peoples' Movement*. Cambridge: Cambridge University Press. https://doi.org/10.1017/S0010417505000241.

"Nittak Hullo Chito Na Yupka." 2021. Choctaw Nation of Oklahoma. December 2021.

Noble, Safiya Umoja. 2018. *Algorithms of Oppression: How Search Engines Reinforce Racism*. New York: New York University Press.

NPS (National Parks Service). 2023. "Indigenous Fire Practices Shape Our Land." September 5, 2023. https://www.nps.gov/subjects/fire/indigenous-fire-practices-shape-our-land.htm.

O'Brien, Greg. 2005. *Choctaws in a Revolutionary Age, 1750–1830*. Lincoln: University of Nebraska Press.

Orgad, Shani. 2009. "How Can Researchers Make Sense of the Issues Involved in Collecting and Interpreting Online and Offline Data?" In *Internet Inquiry: Conversations about Method*, edited by Annette N. Markham and Nancy K. Baym, 33–67. Los Angeles, CA: Sage Publications.

Ortiz, Elena. 2023. "Elena Ortiz on the Fields Museum's *Native Voices* Exhibit." @TheRedNation, YouTube, July 2023. https://www.youtube.com/watch?v=Rt2DATgQmK8.

Parker, Kathleen. 2018. "Elizabeth Warren's DNA Test Is Really Much Ado about Very Little." *Washington Post*, October 16, 2018. https://www.washingtonpost.com/opinions/elizabeth-warrens-dna-test-really-is-much-ado-about-very-little/2018/10/16/0d2a7b8a-d184-11e8-b2d2-f397227b43f0_story.html.

Pesantubbee, Michelene E. 2005. *Choctaw Women in a Chaotic World: The Clash of Cultures in the Colonial Southeast*. Albuquerque: University of New Mexico Press.

Pew Research Center. 2018. "The #MeToo Hashtag Has Been Used Roughly 19 Million Times on Twitter in the Past Year, and Usage Often Surges around News Events." *Pew Research Center*, October 11, 2018. http://www.pewresearch.org/fact-tank/2018/10/11/how-social-media-users-have-discussed-sexual-harassment-since-metoo-went-viral/ft_18-10-11_metooanniversary_hashtag-used-19m_times/.

Phillips, Whitney. 2015. *This Is Why We Can't Have Nice Things: Mapping the Relationship between Online Trolling and Mainstream Culture*. Cambridge, MA: MIT Press.

Pocahontas. 1995. Written by Carl Binder, Susannah Grant, and Philip Lazebnik. Directed by Mike Gabriel and Eric Goldberg. Disney.

Powell, Malea, Daisy Levy, Andrea Riley Mukavetz, Marilee Brooks-Gillies, Maria Novotny, Jennifer Fisch-Ferguson, and Cultural Rhetorics Theory Lab. 2014. "Our Story Begins Here: Constellating Cultural Rhetorics." *Enculturation: A Journal of Rhetoric, Writing, and Culture*, October 25, 2014. http://enculturation.net/our-story-begins-here.

Ramos, Jairo. 2014. "A Native American Tribe Hopes Digital Currency Boosts Its Sovereignty." *Code Switch*, podcast, March 7, 2014. https://www.npr.org/sections

/codeswitch/2014/03/07/287258968/a-native-american-tribe-hopes-digital-currency-boosts-its-sovereignty.

Ratcliffe, Krista. 2005. *Rhetorical Listening: Identification, Gender, Whiteness*. Carbondale: Southern Illinois University Press.

Reilly, Katie. 2018. "A New North Dakota Law Threatened Native American Votes. They Responded by Turning Out in Historic Numbers." *Time*, November 7, 2018. https://time.com/5446971/north-dakota-native-american-turnout/.

Riley Mukavetz, Andrea, and Cindy Tekobbe. 2022. "'If You Don't Want Us There, You Don't Get Us': A Statement on Indigenous Visibility and Reconciliation." *Present Tense: A Journal of Rhetoric in Society* 9 (2). https://www.presenttensejournal.org/volume-9/if-you-dont-want-us-there-you-dont-get-us-a-statement-on-indigenous-visibility-and-reconciliation/.

Royster, Jacqueline Jones, and Gesa E. Kirsch. 2012. *Feminist Rhetorical Practices*. Carbondale: Southern Illinois University Press.

Shifman, Limor. 2013. *Memes in Digital Culture*. Cambridge, MA: MIT Press.

Smith, Linda Tuhiwai. 2012. *Decolonizing Methodologies: Research and Indigenous Peoples*. 2nd ed. London: Zed Books.

Sparby, Derek M. 2017. "Digital Social Media and Aggression: Memetic Rhetoric in 4chan's Collective Identity." *Computers and Composition*, no. 45, 85–97.

Spencer-Wood, Suzanne M. 2016. "Feminist Theorizing of Patriarchal Colonialism, Power Dynamics, and Social Agency Materialized in Colonial Institutions." *International Journal of Historical Archaeology* 20 (3): 477–91.

Sprague, Joey. 2005. *Feminist Methodologies for Critical Researchers*. Walnut Creek, CA: AltaMira.

Suzack, Cheryl, Shari M. Huhndorf, Jeanne Perreault, and Jean Barman, eds. 2010. *Indigenous Women and Feminism: Politics, Activism, Culture*. Vancouver: University of British Columbia Press.

Tekobbe, Cindy. 2015. "Attack of the Fake Geek Girls: Challenging Gendered Harassment and Marginalization in Online Spaces." PhD diss., Arizona State University. https://keep.lib.asu.edu/items/153506.

Tekobbe, Cindy Kay. 2013. A Site for Fresh Eyes: Pinterest's Challenge to 'Traditional' Digital Literacies. *Information, Communication and Society* 16, no. 3 (January): 381–396. https://doi.org/10.1080/1369118X.2012.756052.

Tekobbe, Cindy. 2019. "Listen: Survivance and Decolonialism as Method in Researching Digital Activism." *Second International Handbook of Internet Research*, edited by Jeremy Hunsinger, Matthew M. Allen, and Lisbeth Klastrup, 979–93. New York: Springer.

"Spatial." 2016. TEMPLATED. Accessed April 1, 2016. http://www.templated.co/spatial.

Tiidenberg, Katrin. 2013. "Online Flashers? Arousal or Offense upon Receiving Penis-Pictures from Audience in Self-Shooters Community." Conference presentation for Console-ing Passions, Leicester, UK, June 23, 2013.

Tiidenberg, Katrin. 2017. "NSFW on Tumblr." Conference presentation for the Association of Internet Researchers, Tartu, Estonia, October 20, 2017.

Tingle, Tim. 2003. *Walking the Choctaw Road: Stories from Red People Memory*. El Paso, TX: Cinco Puntos.

Townsend, Camilla. 2004. *Pocahontas and the Powhatan Dilemma*. New York: Hill and Wang.

Vizenor, Gerald. 1999. *Manifest Manners: Narratives on Postindian Survivance*. Lincoln: University of Nebraska Press.

Wachter-Boettcher, Sara. 2017. *Technically Wrong: Sexist Apps, Biased Algorithms, and Other Threats of Toxic Tech*. New York: Norton.

Walls, Douglas M., and Stephanie Vie. 2017. *Social Writing / Social Media: Publics, Presentations, and Pedagogies*. Fort Collins, CO: WAC Clearinghouse.

Warnick, Barbara, and David S. Heineman. 2012. *Rhetoric Online: The Politics of New Media*. 2nd ed. New York: Peter Lang.

Watanabe, Sundy. 2014. "Critical Storying: Power through Survivance and Rhetorical Sovereignty." *Counterpoints* 449:153–170. https://www.jstor.org/stable/42982070.

Whitebear, Luhui. 2021. "2020 & the Elections Can't Stop Us: Hashtagging Change through Indigenous Activism." *Spark: A 4C4Equality Journal*, no. 3 (2021). https://sparkactivism.com/volume-3-call/hashtagging-change-through-indigenous-activism/.

Wieser-Weryackwe, Kimberly. 2023. "Aunt Ruby's Little Sister Dances." *Unpapered: Writers Consider Native American Identity and Cultural Belonging*, edited by Diane Glancy and Linda Rodriguez, 191–208. Lincoln: University of Nebraska Press.

Williams, Mandy. 2018. "Is MazaCoin, the Native Cryptocurrency, Gaining Popularity Again?" *CryptoPotato*, May 13, 2018. https://cryptopotato.com/is-mazacoin-the-native-cryptocurrency-gaining-popularity-again/.

Wilson, Shawn. 2008. *Research Is Ceremony: Indigenous Research Methods*. Winnipeg: Fernwood Publishing.

Younging, Gregory. 2018. *Elements of Indigenous Style: A Guide for Writing by and about Indigenous Peoples*. Edmonton: Brush Education.

Zimmer, Carl. 2018. "Before Arguing about DNA Tests, Learn the Science behind Them." *New York Times*, October 18, 2018. https://www.nytimes.com/2018/10/18/opinion/sunday/dna-elizabeth-warren.html.

Index

abuse, online, 68–69
academic programming, Indigenous, 158
accountability, in research, 165–66
activism, 53, 55, 56, 66, 73; Indigenous, 28, 34, 135, 161; public messaging, 162–63
Adams, Mikaëla, *Who Belongs?*, 49
Aila test, 21
AIM. *See* American Indian Movement
alcoholism, as stereotype, 93–95
alt-currencies, alt-coin, 30, 131, 135. *See also* cryptocurrency; MazaCoin
American Indian Caucus (CCCC), 113
American Indian Movement (AIM), 27
American Indian rights movements, 27
American Rescue Plan, broadband access, 156–57
ancestors, repatriation of, 156
antigovernment rhetoric, 147
Arizona, voter suppression in, 76
Arola, Kristin, "Indigenous Interfaces," 83–84
assimilation, and generational trauma, 27
Association of Internet Researchers, 136
audience: cultural sharing, 56–57; MazaCoin, 148; social media, 147
"Aunt Ruby's Little Sister Dances" (Wieser-Weryackwe), 48–49

Austin, Jack, Jr., 3
authenticity, of Payu Harris, 141–42
autonomy, 70, 143; losses of, 147–48

ballot collection, 76
Barlow, John, "A Declaration of the Independence of Cyberspace," 44, 145
Batman, 29, 107
"Battle of Little Bitcoin: Native American Tribe Launches Its Own Cryptocurrency, The" (Hamill), 139–40, 148–49
Batton, Gary, 3
bears: grizzly, 99, 100; in stories, 106–7
BIA. *See* Bureau of Indian Affairs
Biden, Joe, 21, 156
Biskinik, The (newspaper), 4, 38
Bitcoin, 131, 161; and user class, 144–45
Bitcoin Oyate Project (BTC Oyate), 138, 142
blood quantum, 13, 170n7; and identity, 47–48, 88–89, 101–2; tribal membership and, 48–49
boarding schools, 60, 156
Boxley, David, 113
Bradbury, Danny, 135, 147; "MazaCoin Aims to Be Sovereign Altcoin for Native Americans," 138–39
Brayboy, Bryan, 36

broadband access, 156–57
BTC Oyate. *See* Bitcoin Oyate Project
Buck, Amber, 14, 154
Build Back Better, broadband access, 156–57
building connection, 110–11
Bureau of Indian Affairs (BIA), 76
Burke, Tarana, #MeToo, 63

Canada, 156, 163
capitalism, 48, 83, 86, 132, 157; Pine Ridge Indian reservation, 29–30; settler-colonialism, 91, 161
Carlisle Indian School, 156
ceremony, 19
change-making, 73
Cherokee identity, 48
Choctaw (Chahta), 6, 9, 13, 36, 113, 169n1; gift of corn, 3, 4–5; identity, 48, 49; matriarchy vs. patriarchy and, 59–60
Choctaw Nation, 14, 163; blood quantum and membership, 13, 48; Christmas cards, 3–4
Choctaw Nation YouTube, 113
Christianity: Evangelical, 61; influence of, 4, 120; and social media, 124–25
cishetmale, 66–67, 68; sexual harassment, 72–73
Civil Rights movement, 27, 47
civil unrest, 27
climate change, and water, 162
code talkers, Navajo, 20, 21
Coindesk, 138
collaboration, 9, 18, 26, 60, 157, 165; social media and, 121–22; in storytelling, 53–54
collection, of memes, 46–47
collectivity, 84, 165; and MazaCoin, 140, 148–49; memory, 41; responsibility, 67
colonization, colonialism, 20, 24, 70, 132, 150. *See also* settler-colonialism
comic book art, 28–29, 107
communalism, promoting, 148–49
communities, 35–36
community-building, 81
Conference on College Composition and Communication Native American Special Interest Group, 158
connection, through identification, 110–11
Consunji, Blanca, 142
Contemporary Feminist Research from Theory to Practice (Leavy and Harris), 108
corn (*Zea mays*), Choctaw origin stories of, 3, 4–5, 37–38, 40
Corn Lady; Corn Goddess (Ohoyo Osh Chishba), 3, 4–5, 37–38, 40

Cottom, Tressie McMillan, 25–26
COVID-19, 154; and economies, 30–31; responses to, 119, 120
Creepingbear, Shane, 44–45
Critical Indigenous Research Methodologies (CRIM), 13–14
critical race theory, 19
"Critical Storytelling: Power through Survivance and Rhetorical Sovereignty" (Watanabe), 13
criticism, 115–16; as artist, 125–27
cryptocurrencies, 15, 16–17, 45–46, 130, 133, 137, 146, 150; MazaCoin and, 29–30, 131; and Oglala Sioux economy, 141–42; social values of, 144–45
culture(s), 11; digital, 157; learning, 113–14; practices, 9, 68; retention, 112–13; sharing, 56–57
culture-making, 50
cyberspace, regulation of, 145

Dakota Access Pipeline (DAPL), 23, 28, 81, 91
data, monetization of, 42
Dawes Commission/Rolls, 49, 169–70n6, 170n9; and blood quantum, 47–48, 88–89
"Declaration of the Independence of Cyberspace, A" (Barlow), 44, 145
decolonization, decolonialism, 46, 53, 68, 105, 158, 161, 164, 166; and neoliberalism, 134–35; of research methodology, 108–9; of self, 155, 157; of sexual harassment/assault, 65–66
Decolonizing Methodologies (Smith), 54, 137, 167
DEI. *See* diversity and inclusivity initiatives
dick pics, 69, 72
digital platforms, 19, 42; Veregge on, 120–25
Digitizing Race (Nakamura), 77
disenfranchisement, 16, 76
Disney, Pocahontas, 82
distance, objective, 137
diversity and inclusivity (DEI) initiatives, 158
DNA testing, and identity, 101–2

Ecoffey, Brandon, "Oglala Sioux Tribe Surprised by MazaCoin Plan," 141
economies: COVID-19 and, 20–31; neoliberal, 43, 61; Pine Ridge Reservation, 130, 140–41
education, 4, 60, 166; on structural bias, 64–65; Jeffrey Veregge's, 111–12
Elements of Indigenous Style (Younging), 54
elites, technological, 144–45
emotion, and political memes, 28
Engel, Evan, 142
enrollment, 14, 49
environmentalism, 162; as concern, 119–20

"Epic Superhero Art in a Traditional Native American Style" (Newitz), 121
epistemologies, Indigenous, 7, 25
ethics, research, 19
Ethics of Internet Research, The (McKee and Porter), 42
Evangelical Christianity, 61

Facebook, 27, 79, 80, 82; identity making in, 42–43; memes, 46–47; settler-state codification in, 44–45
faking it, 112–13
family, 12, 14, 16, 60, 163–64, 170n3
females, film representations of, 21. *See also* Pocahontas
feminisms, 159; Indigenous, 99–101; intersectional, 25, 53, 60–61; white, 63, 81, 97–98
feminist research/theory, 8–9
films, female Indigenous characters in, 21
financial technologies, and user classes, 144–45
flashing, 70, 72
Fox, 6, 26
frameworks, Indigenous, 25

Gawker Media LLC, 71
Gee, James Paul, 36
gender, 66–67; in Indigenous and Western cultures, 98–99
gender theory, 19
genocide, 27
Gizmodo, 29
Gizmodo Media Group, 71
Glisic, Albina, 97
good relations, 6–7, 19, 41, 56, 160
governance, settler-state, 70
graves, at boarding schools, 156
Great Spirit, 4
Greenfield, Adam, *Radical Technologies*, 144
Grim, John A., "Indigenous Lifeways and Knowing the World," 83
grizzly bears, 99, 100

Hackett, Isa, 62
Hamill, Jasper, "The Battle of Little Bitcoin," 139–40, 148–49
Harris, Anne, *Contemporary Feminist Research from Theory to Practice*, 108
Harris, Kamala, 21
Harris, Payu, 43; authenticity of, 141–42; communalism, 148–49; digital journalist coverage of, 131–32; cryptocurrency, 15, 16–17, 130; identity, 44, 45, 46; and MazaCoin, 29–30, 134, 138–39, 161; media characterization of, 142–43; storytelling and thick meaning, 139–40
Heineman, David S., *Rhetoric Online*, 42
"Here's How America Uses Its Land," 89; meme, 90–91
heteropatriarchy, 20
humor, 79
Hunter and the Alligator, The, 32–34, 40
hypercapitalism, 137

identification, connection through, 110–11
identity, identities, 7, 16, 22, 42, 47, 73, 83, 118, 119, 142, 160; blood quantum and, 13, 48, 88–89, 101–2; changing, 8–9; digital, 50–51, 164; government-sanctioned, 49–50; Indigenous, 15, 28, 155; intersectionality, 60–61; political, 14, 81, 84; remaking, 9–12; thickness of, 24, 44; Jeffrey Veregge's, 109–10
identity construction, identity-making, 18, 36, 41, 46, 50, 159; digital, 77, 77–78; on Facebook, 42–43; memes and, 78–79
identity practices, 26, 80, 118
identity work, 7–9, 45, 135, 161
Idle No More (#IdleNoMore) movement, 23, 28
"'If You Don't Want Us There, You Don't Get Us': A Statement on Indigenous Visibility and Reconciliation" (Mukavetz), 155
incarceration, Indigenous rates of, 76–77
"*Inception*" movement, 29
Indian Citizenship Act, 76
Indian Country, criticism from, 126–27
Indian Country Today, on MazaCoin, 140–41
Indianness, 13, 14, 18, 30, 114
Indian Removal, 49
Indian Removal Act, 81, 157; and blood quantum, 47–48
indigeneity, 10–11, 15, 111, 155; and self-identification, 12–13
"Indigenous Interfaces" (Arola), 83–84
"Indigenous Lifeways and Knowing the World" (Grim), 83
Indigenous practices, 53, 67
"Indigenous Women" meme, 99–100
individualism, 18, 83, 118
Intersectional Internet (Nobel and Tynes), 77
intersectionality, 81; feminisms, 35, 60–62
intertextuality, 78
interviewing, interview, 112; building connection, 110–11; process, 109–10; public voice and, 117–19; as research method, 108–9
introduction, ritual of, 110–12

io9 website, 121
isolation, stereotype of, 134

Jezebel media site, 70–71
journalism, 17, 20, 30; digital, 131–32
Judd, Ashley, 62

Kaepernick, Colin, 74
kairos, 54
King, Thomas, on stories, 36–37
kinship networks, 34, 35, 81
knowledge(s), 7, 19, 81, 111; access to traditional, 112–13; gaining, 5–6; Indigenous, 100, 157, 159; production of, 6, 34, 43; western, 45, 62
knowledge-making, 158–59, 160
knowledge practices, knowing, 9, 35, 43; Indigenous, 5, 11, 55–56, 80–81; thick, 25–26
Know Your Meme, 93
Kovach, Margaret, 40–41
Kowi Anukasha (Forest Dwellers; Little People), 129–30

"Lakota Nation Adopts Mazacoin Cryptocurrency as Legal Tender" (*RT News*), 148
land, land rights, 77, 84, 95; allotment, 47–48; memes on, 89–93; repurchasing ancestral, 149–50; stolen, 96–97
#LandBack movement, 91
Landry, Alysa, "9 Questions Surrounding MazaCoin, the Lakota CryptoCurrency: Answered," 140–41
language(s), 24, 58
Last Real Indians website, 142
Leavy, Patricia, *Contemporary Feminist Research from Theory to Practice*, 108
Leonard, Michele, "You Don't Look Indian," 13
line art, Jeffrey Veregge's, 107, 112–13
Little Bighorn, Battle of, reference to, 148, 149
Littlefeather, Sacheen, 16
Little People, The, 129–30
Lone Hill, Dana, 142, 171n3
Lovink, Geert, 10

mainstream media, on Native Americans, 133–34
Manifest Destiny, 83, 95
marginalization, 42, 155, 165
Mashable, 142
material assets, and race, 48
matriarchy, vs. patriarchy, 59–60
Maza, 142–43

MazaCoin, 16–17, 29–30, 43, 45, 148, 149–50; and Pine Ridge economy, 140–41; publicity on, 131–32; research on, 133, 134–35; technologies and narratives of, 145–46; website, 142, 146–47
"MazaCoin Aims to Be Sovereign Altcoin for Native Americans" (Bradbury), 138–39
McClintock, Tom, 100
McKee, Heidi A., *The Ethics of Internet Research*, 42
McKnight, John Carter, 16, 17, 30; and cryptocurrency project, 45–46, 135, 136–37, 141
meaning(s): of stories, 5, 39, 40; thickness of, 24, 53
meaning-making, 26, 41–42; storytelling and, 55–56
media: on Payu Harris, 142, 143; mainstream, 133–34; on Oglala Lakota, 135–36
memes, 36; Facebook, 46–47; identity and affinity constructions, 78–79; Indigenous political, 27–28, 43, 160; "Indigenous Women," 99–100; on land issues, 89–93; "Obama and Trump Indian Policy," 81–82; Pocahontas, 86–89; settler-colonialism, 84–86; "Settlers: We Have Culture," 96–97; "Three Minutes Later," 93–95; "True Spirit of Thanksgiving," 103–4; Elizabeth Warren in, 101–2
mentoring, 112–13
methodologies, 7, 60, 160; decolonizing, 108–9; digital research, 80–81; scientific classifications, 39–40; storytelling as, 18–19, 26–27. *See also* Critical Indigenous Research Methodologies
#MeToo movement, 26–27, 43, 72, 159–60; cultural impact, 54–55; intersectionalities of, 61–62; personal growth/development, 63–64; social media spread of, 62–63; structural bias and, 64–65
Milano, Alyssa, #MeToo movement, 54, 62–63
Missing and Murdered Indigenous Women (#MMIW, #MMIWG, #MMIWC), 21, 23, 28, 77, 163
Montana, 76
Mukavetz, Andrea Riley, "'If You Don't Want Us There, You Don't Get Us,'" 155
multiculturalism, 50, 164–65
murals, Jeffrey Veregge's, 107

Nahdee, Ali, 21
Nakamoto, Satoshi, 144
Nakamura, Lisa, *Digitizing Race*, 77
names, Native American, 44–45

nationalism, 22
Native American Graves Protection and Repatriation Act (NAGPRA), 163
Native TikTok, 155
Navajo Nation, 21
neocolonialism, as neoliberalism, 73–74
neoliberalism, 11, 43, 44, 53, 61; and decolonization, 134–35; and neocolonialism, 73–74; and privilege, 67–68
nerd culture art, 29
networks: building, 111; digital, 120–21
network theory, 34
Newitz, Annalee, 107
news stories, Native Americans in, 133–34
New York Bitcoin Center, 140
"9 Questions Surrounding MazaCoin, the Lakota CryptoCurrency: Answered" (Landry), 140–41
Noble, Safiya, 44; *Intersectional Internet*, 77
#NoDAPL campaign, 23, 28, 47
nonbinary women, 100
North Dakota, voter suppression, 22–23, 76

Octopus Woman sculpture (Veregge), 116–17
Obama, Barrack, NODAPL protest, 47
"Obama and Trump Indian Policy" meme, 81–82; themes in, 91–93
Oglala Lakota, 44, 45, 130, 131, 143; and cryptocurrency, 141–42; MazaCoin and, 29–30; and Maza website, 146–47; media depictions of, 135–36, 148–49; poverty, 138, 139
"Oglala Sioux Tribe Surprised by MazaCoin Plan" (Ecoffey), 141–42
Ohoyo Chisba Osh / Ohoyo Osh Chishba (Unknown Woman), 3; story variations, 4–5, 37–38, 40
Orgad, Shani, 43
Ortiz, Elena, 7
Owl Woman, The, 151–53

Pacific Northwest, line art, 107
pandemic, and Zoom, 109
patriarchy, 61, 62, 70, 73; vs. matriarchy, 59–60
peer-to-peer finance, 145
personhood, settler-state, 70
phenotype, and race, 50
physical addresses, and voter suppression, 76
Pine Ridge Indian Reservation, 130, 140; digital journalists on, 131–32; MazaCoin, 17, 29–30, 46; poverty, 138–39
PO boxes, and voter suppression, 76
Pocahontas, 82–83; memes, 86–89, 105; stereotypical/racist images of, 21, 22

podcasts, Indigenous, 163
politics, 105, 123, 160
pop culture art, 29
pornography, social media and, 69, 70–71, 72
Porter, James E., *The Ethics of Internet Research*, 42
Port Gamble reservation, 112
poverty, Pine Ridge Reservation, 138–39
pretendians, 16, 22, 50
Price, Roy, 62
prison-industrial complex, 76–77
privilege, 65; neoliberalism and, 67–68
protests, Civil Rights movement, 47
provocation, 66
public awareness, 166
public discourse, on repatriation, 156
public voices: Indianness and, 114; Jeffrey Veregge's, 117–19

rabbit, in stories, 75, 106–7
race, 48, 50
racism, 23, 47, 89; stereotypical identities, 15, 171n3; Trump's, 20–21, 22
Radical Technologies (Greenfield), 144
rape culture: decolonization of, 65–66; social media and, 70–71, 72
reality, 6, 59; and worldview, 40–41
reciprocity, 132
Reddit, and MazaCoin, 142, 149–50
Red Wolf (Veregge), 171n6; criticism of, 126–27
relational networks, 34, 35
relations, 35, 57, 59, 67, 100, 118, 162
relationships, 111, 132; Indigenous knowledge, 80–81; with social media, 154–55
religion, 61
repatriation, 156
research, researcher, 5–6, 57, 133, 143; accountability, 165–66; digital, 80–81
Research Is Ceremony: Indigenous Research Methods (Wilson), 6, 41–42
Reservation Dogs, 13
resistance, 56, 77, 86, 150, 158; decolonization, 68, 155
Rezzy Red Proletariat Memes, 27–28, 88; on land issues, 89–91; methodologies, 80–81
Rhetoric Online: The Politics of New Media (Warnick and Heineman), 42
rhetorics: cultural, 5, 7, 9, 35, 36, 53, 59, 149, 165; digital, 19, 23
Roanhorse, Rebecca, 107
Russia Today (RT News), 171n4; on MazaCoin, 134, 148

sacredness, 124, 162
satire, 79
schools, 156
science: classification in, 39–40; and traditional knowledge, 100
science fiction and fantasy, 114–15; Jeffrey Veregge's art, 107–8
sculpture, Veregge's, 116–17
"Seduction of the Moon" (Veregge), 116
self, thick meaning of, 127–28
self-determination, 56, 70
self-identification, self-identity, 8, 9, 10, 53, 104; indigeneity and, 12–13
self-location, 56
self-promotion, 44
self-reflection, 8, 9
settler-capitalism, 8, 25
settler-colonialism, 4, 20, 24–25, 91, 119, 156, 160, 162; artistic criticism form, 125–26; in memes, 85–86; Pocahantas images, 82, 83
"Settler-Colonialism" meme, 97–98
settler logic, 59
settlers, 83; land and water rights, 92–93
"Settlers: We Have Culture" meme, 96–97
settler-state, 44, 60, 70
sexism, Trump's, 20–21
sexual abuse, 69, 170n2
sexual assault, 62, 69; decolonization of, 65–66
sexual harassment, 62, 62; decolonization, 65–66; personal growth and development, 63–64; social media and, 69–72; storytelling and, 72–73
sexualization, of Indigenous women, 21, 82–83
Shilup Chitoh Osh (The Great Spirit of the Choctaws), 38
S'Klallam tribe, 110
Smith, Adam, 86
Smith, John, 82
Smith, Linda Tuhiwai, 39, 111; *Decolonizing Methodologies*, 54, 137, 167
social inequity, 60
social justice, 74, 124
social media, 18, 19, 41, 42, 43, 44, 66, 83, 104, 105, 148; Maza, 146–47; #MeToo movement, 62–63; relationships with, 154–55; sexual harassment and abuse, 69, 70–72; Veregge on, 120–25
social networking, 43
social values, cryptocurrencies, 144–45
South, Indigenous identity in, 49
sovereignty, 14, 23, 169n2
space/place, digital, 17, 23–24, 164; Indigenous vs. western, 137–38

Spanos, Nick, 140
spirituality, 120; in social media, 123–25
#StandingRock, 28
Standing Rock protests, 81, 82, 84, 91
stereotypes, 13, 15, 17, 20, 21, 24–25, 134; alcoholism, 93–95; and public voice, 118–19
storytelling, stories, 3, 4–5, 9, 16, 39, 45, 78, 115, 140, 158; collaborative, 53–54; "Corn—A Choctaw Legend," 37–38; *The Hunter and the Alligator*, 32–34; *The Little People*, 129–30; and meaning-making, 55–56; as methodology, 18–19, 26–27; *The Owl Woman*, 151–53; role/purpose of, 36–37; and sexual harassment, 72–73; *Why the Rabbit Has a Short Tail*, 75; *Why the Turtle Has Cracks on His Back*, 52–53; *Why Rabbit Is So Lean*, 106–7; and worldview, 40–41
structural bias, 64–65
superheroes, Jeffrey Veregge's version of, 107
survivance, 53, 73; and embodied practice, 57–58
Survivance (Vizenor), 56

taxonomy, as hierarchy, 44
technology: digital, 18; as ideological, 44
technology industry, 8; culture of, 11, 42; financial, 144–45
technolibertarianism, 44
Thanksgiving, 104
thick meaning, 24, 25–26, 53, 146; of self, 127–28
"Three Minutes Later" meme, 93–95
"Three Drinks Later," 93
Tiidenberg, Katrin, 69
time, Indigenous vs. western, 137–38
time travel, 114–15
trauma, 95; generational, 27
treaties, and land rights, 95
tribal critical race theory (TribalCrit), 65
tribe(s): identification cards, 23, 76; membership, 48–49; and self, 128
trolls, sexual harassment by, 70–72
"True Spirit of Thanksgiving" meme, 103–4
Trump, Donald, 43, 47, 55, 88, 91, 170n2; racism and sexism, 20–21, 22; and Elizabeth Warren, 87, 170n8, 170n10
Truth and Reconciliation Project, 163
Tsimshian style, 113
Turtle, 52–53
Turtle Island, 39, 169n3
2020 concerns, 119–20, 125
Twitter, 43
Tynes, Brendesha, *Intersectional Internet*, 77

unhoused populations, voter suppression and, 77
US Supreme Court, voter suppression, 22–23
University of Alabama, repatriation and, 156
Unknown Woman. *See* Ohoyo Osh Chishba Osh
Unpapered (Glancy and Rodriguez), 16
use handles, 10
user classes, and cryptocurrencies, 144

Veregge, Jeffrey, 14, 28–29, 43–44, 107, 160–61; as artist, 125–26, 128; criticism of, 115, 125–27; on cultural retention, 112–13; on digital platforms, 120–25; interview process, 109–10, 114; introduction of, 111–12; Octopus Woman sculpture, 116–17; public voice, 117–19; "Seduction of the Moon," 116; 2020 concerns, 119–20
victimization, 21
violence, 37, 47
Vizenor, Gerald, *Survivance*, 56
voter suppression, 22–23; prison-industrial complex and, 76–77
voting, access to, 76

Wachter-Boettcher, Sara, 44
Warnick, Barbara, *Rhetoric Online*, 42
Warren, Elizbeth, 16, 81, 82; in memes, 101–2; Native ancestry story, 88–89; Trump and, 20–21, 22, 87–88, 170n8, 170n10

warrior trope, 148, 149
water, 161–62; protecting, 91–93
Water Is Life (#waterislife) movement, 120, 161–62
Water Protectors, 84, 161–62; role of, 91–93
websites, 78; Choctaw Nation, 163; cryptocurrency, 133, 146–47
Weinstein, Harvey, 62
"when you see the 'invisible hand'" meme, 84–866
whiteness, 14–15, 158
white supremacy, 15, 20, 44, 45, 47, 164; and Indigenous identity, 49, 50
Who Belongs? Race, Resources, and Tribal Citizenship in the Native South (Adams), 49
Why Rabbit Is So Lean, 106–7
Why the Rabbit Has a Short Tail, 75
Why the Turtle Has Cracks on His Back, 52–53
Wieser-Weryackwe, Kimberly, "Aunt Ruby's Little Sister Dances," 48–49
Wilson, Shawn, 79; *Research Is Ceremony*, 6, 41–42
women, 21; in Indigenous cultures, 98–100
worldview, 40–41

"You Don't Look Indian" (Leonard), 13
Younging, Gregory, *Elements of Indigenous Style*, 54

Zoom, pandemic and, 109

About the Author

Cindy Tekobbe is a scholar of critical feminist science and technology, dual-appointed in Gender and Women's Studies and Communication, at the University of Illinois Chicago. Her work investigates the digital lives, identities, and activism practices of traditionally underrepresented and erased peoples and communities. She is enrolled with the Choctaw Nation of Oklahoma.